Proprioceptive neuromuscular facilitation

Patterns and techniques

Proprioceptive neuromuscular facilitation

Patterns and techniques

Second edition

MARGARET KNOTT, B.S.

Coordinator of Patient Services, Kaiser Foundation
Rehabilitation Center, Vallejo, California

DOROTHY E. VOSS, B.Ed.

Associate Professor of Physical Therapy, Department
of Rehabilitation Medicine, Northwestern University
Medical School, Chicago, Illinois

WITH ILLUSTRATIONS BY HELEN DREW HIPSHMAN AND JAMES B. BUCKLEY

foreword by Sedgwick Mead, M.D.

Medical Department
Harper & Row, Publishers
New York, San Francisco, London

The highest happiness of man as a thinking being
is to have probed what is knowable
and quietly to revere what is unknowable.

GOETHE, *Maxims and Reflections*

SECOND EDITION / PROPRIOCEPTIVE NEUROMUSCULAR FACILITATION /
PATTERNS AND TECHNIQUES /
*Copyright © 1968 by Hoeber Medical Division; First Edition © 1956 by Hoeber Medical
Division, Harper & Row, Publishers, Incorporated. Printed in the United States of America.
All rights reserved. For information address Medical Division, Harper & Row, Publishers, Inc.
2350 Virginia Ave., Hagerstown, MD 21740*

LIBRARY OF CONGRESS CATALOG CARD NUMBER: *68–21956*

Contents

List of illustrations

List of tables

Foreword

I welcome this new edition of a work which is now a classic in both kinesiology and treatment. My personal experience with the method goes back to 1954. Initially skeptical about it, I have come to feel that it exceeds all other methods of therapeutic exercise in speed of improvement, economy of time, and thoroughness of results.

The spiral-diagonal patterns were observed by Dr. Kabat as a fact of natural movement in healthy human subjects. His description is empirical and not guaranteed to be complete. Years of experience have brought about few changes, however.

Physicians often profess to be baffled by proprioceptive neuromuscular facilitation patterns and techniques, and indeed they cannot be learned without a great deal of study and demonstration. Few physicians have been willing to take the requisite amount of time. Some are tempted to skim through the book, skim through a teaching session, and then pronounce a judgment.

The serious student will find the new revision rewarding. Those with no prior acquaintance with the basic premises or methods would do well not to try learning them by home correspondence alone.

SEDGWICK MEAD, M.D.
Medical Director,
Kaiser Foundation Rehabilitation Center,
Vallejo, California

Preface

More than a decade has passed since the First Edition was published. Responses to the First Edition have ranged from "I can't even read it, let alone understand it!" to "I find the book *most* helpful." Hopefully, because this material is now a part of the curricula of many physical therapy schools, responses to this new edition will reveal that frustration may have lessened.

Interest around the world has continued to grow as evidenced by the almost three hundred graduate physical therapists from other countries who have come to learn, and return. More than two hundred in the United States have participated in short-term courses of several weeks' duration and some have devoted three or six months to learning. Short-term courses have been given in Argentina, Australia, Canada, Denmark, England, Finland, Guatemala, Norway, and Sweden, and perhaps in other countries. A translated edition appeared first in Germany in 1962. Preparation of a Japanese edition was begun in 1966, and of a translation to French in 1968.

More than two decades have passed since Dr. Herman Kabat began his work in Washington, D.C. The original concepts expressed by Dr. Kabat have been extended far beyond his "treatment for paralysis." To those of us who have applied his ideas more broadly, this approach adds to the understanding of human movement and constitutes a total approach to the area of treatment known as "therapeutic exercise." Somewhere, sometime, we anticipate that application of the material on developmental activities in this new edition will be made in the promotion of motor learning in "normal" children, in the field of physical education, and in the prevention of injury among athletes and workers.

This Second Edition presents aspects of developing motor behavior which seem to provide the base for our use of developmental activities. Here we have relied on the works of Hooker, Gesell and his colleagues, and of McGraw. This material reflects our belief that recapitulation of the sequence of developmental activities has value for all patients; its use should not be limited to the treatment of the patient having cerebral palsy. The sequence permits treatment by design rather than by haphazard, chance decision. The First Edition included the "how to" for the individual patterns. We have now added the "how to" of combining these patterns so as to teach the patient to perform total patterns such as rolling, rising to sit, to stand, and to walk. Guidance is given, too, on the training of selected transfer and self-care activities.

The First Edition was written from our understanding of the responses of normal subjects and from experience with the method; there was no documentation. We have now risked the hazard of being dubbed "amateur neurophysiologists." Even so, documentation in this Second Edition is sparse in many instances. We have relied upon Dr. Kabat's knowledge of the works of Sherrington, but we recommend to all physical therapists the wondrous experience of reading Sherrington's writing. This great man exemplified for all of us the need to understand the components in order to understand the whole without losing sight of the wholeness.

In this Second Edition, all of the procedures suggested for the facilitation of total patterns have a common purpose—to promote motor learning. Oddly, this term strikes some physical therapists as new or foreign, yet we have always tried to "teach the patient" to perform a motor act and have been pleased when "the patient has learned." Again, our understanding of motor learning must be based on work in other fields. We must look to neurophysiology for knowledge of basic mechanisms, to experimental psychology, and to cybernetics and tracking.

Physical therapists who wish to have clear-cut guidelines on the treatment of specific disabilities such as hemiplegia, cerebral palsy, and various orthopedic conditions will be disappointed in this edition. Treatment must be designed for the individual patient, and the design must be based upon general principles and an understanding of normal motor behavior. We have included in the Suggested Readings a list of "Clinical Applications," many of which are case reports, which may be of help. They cannot, however, provide a recipe for the treatment of any patient.

Unfortunatelly, imparting a skill through the written word is difficult. Physical therapists who really want to learn the method must rigorously develop the necessary skill, with the normal subject serving as the best possible "equipment." The learner needs the balanced antagonism, the resiliency and bounce of the normal subject. The pitfall inherent in trying to learn by working with patients is that, by virtue of previous education and practice, the physical therapist becomes involved with the periphery or focuses on the *most* severely involved segment or muscle group. Proprioceptive neuromuscular facilitation enlists the *less* involved parts, to promote a balanced antagonism of reflex activity, of muscle groups, and of components of motion. Those who cannot resist the temptation to work with patients before they have acquired understanding and have practiced with normal subjects would do better by beginning with total patterns as presented in mat activities.

The contributions of others to this book are noteworthy. The illustrations for the First Edition, drawings by Helen Drew Hipshman of San Francisco, have been retained with minor changes which seemed necessary for consistency. The new drawings of total patterns were done by James B. Buckley of Chicago. Three physical therapists—Margaret Hennessy of British Columbia, Inge Berlin of Germany and Lorna Brand of Wales—served as subjects for photographs of total patterns. Carl Manner of Vallejo was the photographer. The time and energy contributed by these four persons made it possible for Mr. Buckley to produce the new drawings.

Two patients known to us most generously gave financial support for the new illustrations. While they must remain unnamed, we are deeply grateful to them.

A special word of appreciation is owed to Dr. Mead. His continuing interest and support have helped in many ways, ways that are important to patients and to physical therapists.

To our publisher, especially to Paul B. Hoeber who has maintained his interest and enthusiasm, since 1954, and to Claire Drullard, who worked with us on this new edition, we offer gratitude for their patience and effort in our behalf.

In a sense, every graduate physical therapist, every student, and every patient with whom we have worked is a part of this book. As we hope they have learned from us, we have learned from them. The learning process is in continuum.

M.K.
D.E.V.

Introduction

Before presenting an approach to neuromuscular education, reeducation, or therapeutic exercise, it is well to consider its underlying philosophy. Since living human beings are subjected to treatment, the philosophy of the treatment is based upon certain truisms of life common to all human beings.

Life, in one sense, may be regarded as a series of responses to a series of demands. The normal living body is an efficient mechanism capable of performing motor activities ranging through a wide scope of skill, power, and endurance. It is limited only by the anatomical structure and inherent and previously learned neuromuscular responses. Where there is deficiency of the neuromuscular mechanism, the individual is unable to respond adequately to the demands of life. Techniques of proprioceptive neuromuscular facilitation involve placing a demand where a response is desired.

Purposeful movements are basic to a successful life; they are coordinate and directed toward an ultimate goal. In treatment, movements are also purposeful and specific, in order that effort may be channeled toward specific goals. Haphazard, bizarre movements hinder rather than help optimum function.

Ability, strength, and endurance are developed by active participation in life. The application of techniques of proprioceptive neuromuscular facilitation recognizes that hidden potentials may exist, that they are developed by response to a demand, and that frequency or repetition of activity is important to the learning process and to the development of endurance. A change in activity promotes recuperative power and reverses the fatigue factor.

Civilization has developed largely through cooperation rather than through survival of the fittest. This implies that the strong help the weak in a mutual effort. In this approach to treatment, the stronger parts of the body are utilized to stimulate and strengthen the weaker parts.

The philosophy of treatment using techniques of proprioceptive neuromuscular facilitation is, therefore, a philosophy based upon the ideas that all human beings respond in accordance with demand; that existing potentials may be developed more fully; that movements must be specific and directed toward a goal; that activity is necessary to the best development of coordination, strength, and endurance; and that the stronger body parts strengthening weaker parts through cooperation lead toward a goal of optimum function.

HISTORY

Historically, the method of proprioceptive neuromuscular facilitation was developed at the Kabat-Kaiser Institute over a period of five years, 1946–1951. Herman Kabat, M.D., relied upon his knowledge of the monumental work of Sherrington and other neurophysiologists—Coghill, McGraw, and Gesell—on motor development, Hellebrandt on responses of normal adult subjects, and Pavlov on conditioning of reflexes, among others (ref. 19).

Greatest emphasis was placed on the application of maximal resistance throughout the range of motion, using many combinations of motions which were related to primitive patterns and the employment of postural and righting reflexes. These motions allowed for two component actions of muscles as well as permitting action to occur at two or more joints. For example, the peroneals were allowed to contract in plantar flexion and eversion instead of straight eversion, and the anterior tibial was stimulated in combination with hip and knee flexion. Positioning of a part was considered valuable in that it helped to obtain a stronger contraction in the desired muscle groups. Motion was performed first in the strongest part of the range, progressing toward the weaker parts of the range of motion. Stretch was applied to groups of muscles, usually synergists, for greater proprioceptive stimulation.

The process of overflow, referred to as reinforcement, was used throughout in whatever combination of motions achieved the desired response. The technique of repeated contractions was used to gain range as well as to improve endurance. Stimulation of many types of reflexes was incorporated in treatment programs.

These methods were used for several years. In 1949, a valuable contribution was made when it was learned that having a patient contract isometrically the agonist, then the antagonist, resulted in increased response of the agonist. Upon evaluation, it was realized that Sherrington's law of successive induction had a definite place in techniques of facilitation. This technique was named "rhythmic stabilization." Soon, following the use of "rhythmic stabilization," it was found that applying the same procedure of alternating resistance to isotonic contractions of antagonist and agonist also had a facilitating effect. This technique was named "slow reversal."

Early in 1951, the combinations of motion being used were analyzed carefully. It was found that the most effective combinations were those which permitted maximum elongation of related muscle groups so that the stretch reflex could be elicited throughout a "pattern." These patterns were spiral and diagonal in character, and, upon study, their similarity to normal, functional patterns of motions was noted. Since 1951, no other specific techniques have been developed but application has been made in mat, gait, and self-care activities as a means of accelerating the learning process as well as improving strength and balance.

DEFINITIONS

Techniques of proprioceptive neuromuscular facilitation are methods of placing specific demands in order to secure a desired response. Facilitation, by definition, means, "(1) the promotion or hastening of any natural process; the reverse of inhibition. (2) specifically, the effect produced in nerve tissue by the passage of an impulse. The resistance of the nerve is diminished so that a second application of the stimulus evokes the reaction more easily." Proprioceptive means "receiving stimulation within the tissues of the body." Neuromuscular (neuromyal) means, "pertaining to the nerves and muscles" (ref. 2). Therefore, techniques of proprioceptive neuromuscular facilitation may be defined as methods of promoting or hastening the response of the neuromuscular mechanism through stimulation of the proprioceptors.

PRINCIPLES

By definition and demonstration, the techniques of this method are related to normal responses of the neuromuscular mechanism. Knowledge of the normal neuromuscular mechanism, including motor development, anatomy, neurophysiology, and kinesiology, is basic to the learning of the method. Knowledge of the abilities and limitations of the normal subject from birth to maturity is basic to intelligent application in the treatment of patients who present neuromuscular dysfunction.

1. The normal neuromuscular mechanism is capable of a wide range of motor activities within the limits of the anatomical structure, the developmental level, and inherent and previously learned neuromuscular responses. The innumerable combinations of motion available to the mature, normal subject in meeting the demands of life have been acquired through a well-established developmental pattern and many learning situations requiring physical effort and skill. The normal subject is endowed with reserves of power which may be tapped in stress situations, as evidenced in acts of self-preservation or heroism. The normal subject is also endowed with potentials which may be developed in accordance with environmental influences and voluntary decisions. Extreme evidences of this are the child and octogenarian prodigies.

The normal neuromuscular mechanism becomes integrated and efficient without awareness of individual muscle action, reflex activity, and a multitude of other neurophysiological reactions. Variations occur in relation to coordination, strength, rate of movement, and endurance but these variations do not prevent adequate response to the ordinary demands of life.

2. The deficient neuromuscular mechanism is inadequate to meet the demands of life, in proportion to the degree of the deficiency. Response may be limited as a result of faulty development, trauma, or disease of the nervous or the musculoskeletal systems. Deficiencies present themselves in terms of limitation of movement as evidenced by weakness, incoordination, adaptive shortening or immobility of joints, muscle spasm, or spasticity.

It is the deficient neuromuscular mechanism that becomes the concern of the medical profession and the physical therapist. The medical profession determines the demands which may be placed on the mechanism by the physical therapist in order to develop or redevelop the responses of the neuromuscular mechanism as far as possible. Specific demands placed by the physical therapist have a facilitating effect upon the patient's neuromuscular mechanism. The facilitating effect is the means used by the physical therapist to reverse the limitations of the patient.

PLAN OF VOLUME

Part 1 presents the patterns of facilitation with general information on the combining of components of motion, the line of movement produced by cooperative action of the major muscle components, and relationships between antagonistic patterns. The individual patterns are presented in detail as to motion components and major muscle components. The manual contacts, normal timing, timing for emphasis of specific pivots of action, and the commands are given for each pattern. The illustrations are designed to show the physical therapist's approach to the patient, the manual contacts, the motion characteristics of the patterns through full range, and the motion characteristics of the physical therapist.

The analysis of major muscle components is based upon study of the alignment characteristics of the individual muscles as they are portrayed in the anatomy textbooks, observation, and palpation of muscle action in normal and pathological subjects, and an elementary study determining the position of maximal stretch. The analysis of the position of maximal stretch was performed by placing a suitably sized piece of sheet elastic on the human skeleton at the points of origin and insertion of the individual muscles. The part was then moved from the anatomical position through all possible components of motion—flexion versus extension, adduction versus abduction, external rotation versus internal rotation, or their counterparts of supination and pronation. In the head and neck and trunk, flexion versus extension and rotation with lateral motion toward the same or opposite sides were considered. Because the detailed information of muscle origins and insertions is readily available to all, it is not included in this manual.

*Nomina Anatomica** was reviewed for changes of terminology and a complete revision of terms to be used in this revised edition was considered. There are, however, very few of our previously used terms that would be unintelligible to the person who only

---

* *Nomina Anatomica.* Revised by the International Anatomical Nomenclature Committee of the International Congress of Anatomists, 1950, 1955, 1960. Amsterdam, Excerpta Medica Foundation, 1961.

is familiar with the terms adopted by the International Anatomical Nomenclature Committee. We have accepted *flexor digitorum superficialis* to replace *flexor digitorum sublimus.* Tables 10 to 13, which present optimal patterns for individual muscles, have been revised to include the most recent terminology; these tables may be helpful to some persons.

Part 2 is devoted to the various techniques used to promote the desired response, and to hasten motor learning. An attempt is made to describe the "how to" of the various techniques. It is not possible, within the limitations of a manual, to present in complete detail the application and modifications of the techniques according to specific clinical diagnoses. General knowledge can be applied to a specific situation by an intelligent physical therapist. The summary of techniques presents a partial guide to the selection of techniques, their indications and contraindications. A brief discussion of the use of cold and electrical stimulation as adjunctive agents is included.

Part 3 presents the application of the method in the use of total patterns of movement and posture. Related aspects of normal motor behavior are discussed and an adapted sequence of developmental activities is given. Many of the activities are illustrated with legends which are intended to be helpful to the therapist's learning. Greatest emphasis has been given to mat activities because their use prepares for more advanced activities. Again, the therapist must apply the methods according to the needs of the individual patient.

Part 4 includes application of the method for improvement of vital and related functions. The location of this material does not indicate its relative importance. These functions are of primary importance in the treatment of many patients.

Part 5 presents suggestions for evaluation of the patient's performance and for planning a treatment program.

Part 6 includes references and suggested readings supportive to understanding and application of the method.

Part 7 presents, in tabular form, suggested combinations of patterns for reinforcement, optimal patterns for individual muscles, and a listing of muscles according to peripheral innervation and in relation to individual patterns.

1. Patterns of facilitation

Introduction

The patterns of motion for proprioceptive neuromuscular facilitation are mass movement patterns and are basic in all techniques. Mass movement is a characteristic of normal motor activity and is in keeping with Beevor's axiom that the brain knows nothing of individual muscle action but knows only of movement. In normal, functional motor activity various combinations of motion, or mass movements, require shortening and lengthening reactions of many muscles in varying degrees. Mass movement that is to be a means of placing a specific demand must be a specific combination of motions which is optimum for the specific sequence of muscles primarily responsible for the movement, and it must allow these muscles to contribute their components of action consistently. When performed against resistance, patterns of facilitation promote selective irradiation, a process demonstrated by Sherrington (ref. 18).

The mass movement patterns of facilitation are spiral and diagonal in character and closely resemble the movements used in sports and in work activities. The spiral and diagonal character is in keeping with the spiral and rotatory characteristics of the skeletal system of bones and joints and the ligamentous structures. This type of motion is also in harmony with the topographical alignment of the muscles from origin to insertion and with the structural characteristics of the individual muscles.

There are two diagonals of motion for each of the major parts of the body—the head and neck, upper trunk, lower trunk, and extremities. Each diagonal is made up of two patterns which are antagonistic to each other. Each pattern has a major component of flexion or one of extension, there being two flexion and two extension patterns for each of the major parts. These major components are always combined with two other components.

MOTION COMPONENTS

Each spiral and diagonal pattern is a three-component motion with respect to all of the joints or pivots of action participating in the movement. The three components include flexion or extension, motion toward and across the mid-line or across and away from the mid-line, and rotation.

In describing patterns of facilitation, flexion is always referred to as flexion and extension is always referred to as extension. Motion toward and across the mid-line has its counterpart of adduction with reference to extremity pivots. Motion across and away from the mid-line has its counterpart of abduction. External rotation has its counterparts, supination and inversion. Internal rotation has its counterparts of pronation and eversion.

Head and Neck and Trunk

The patterns of the head and neck and upper trunk are described as flexion or extension with rotation to the left or to the right. The head and neck patterns are the key to the upper trunk patterns and the components of a specific head and neck pattern are continued in the homologous pattern of the upper trunk. The head, neck, and trunk rotate toward the left or right and flexion or extension are combined with motion of the head across the mid-line of the trunk. For example, the pattern of upper trunk flexion to the right has a starting position in which the head, neck, and upper trunk are rotated to the left and hyperextended laterally as though the subject were looking up and above the left shoulder. The pattern proceeds with the head rotating toward the right, the neck flexing

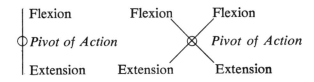

STRAIGHT MOTION   PATTERN OF FACILITATION

and rotating toward the right so that the chin crosses the mid-line of the body as the upper trunk begins to flex with rotation toward the right. The left shoulder approaches the right hip as the motion is completed. The entire motion is one of looking up and above the left shoulder then turning and pulling the head toward the right hip. The directly antagonistic pattern of upper trunk extension with rotation to the left, proceeds from the completed or shortened range of the described motion of upper trunk flexion with rotation to the right.

The lower trunk patterns are described as flexion or extension with rotation to the left or to the right. The bilateral asymmetrical lower extremity patterns are the key to the lower trunk patterns and contribute their respective components of motion. The distal parts of the extremities move across the mid-line of the trunk. Motions of the pelvis include elevation of the pelvic brim as the counterpart of flexion, depression of the pelvic brim as the counterpart of extension, and rotation toward the left or right.

Upper and Lower Extremities

PROXIMAL PIVOTS

The patterns of the extremities are named for the three components of motion occurring at the proximal joints or pivots of action—the shoulder and the hip. Each extremity pattern includes a component of flexion or extension, adduction or abduction, external or internal rotation. There are certain variations between the upper and lower extremity patterns because of the complexity of the upper extremity. In the upper extremity shoulder flexion and extension are combined with adduction and abduction. External rotation is consistent with flexion and internal rotation is consistent with extension. In the lower extremity, hip flexion and extension are combined with adduction and abduction and with external and internal rotation. However, adduction is consistent with external rotation and abduction is consistent with internal rotation.

INTERMEDIATE PIVOTS

The intermediate joints, the elbow and knee, may remain straight or they may flex or extend. The rotation and gliding motions of these joints are consistently in line with the rotation and adduction or abduction occurring at the shoulder or hip. This is true regardless of whether the intermediate action is that of flexion or extension.

DISTAL PIVOTS

Distal components of motion are consistent with proximal components regardless of intermediate joint action. In the upper extremity, supination of the forearm and motion of the wrist toward the radial side are consistent with flexion and external rotation of the shoulder. Pronation and motion of the wrist toward the ulnar side are consistent with extension and internal rotation. Wrist flexion is consistent with shoulder adduction, and wrist extension is consistent with shoulder abduction.

In the lower extremity, plantar flexion of the ankle and foot is consistent with hip extension, and dorsiflexion is consistent with hip flexion. Inversion of the foot and motion toward the tibial side is consistent with hip adduction and external rotation. Eversion of the foot with motion toward the fibular side is consistent with abduction and internal rotation.

DIGITAL PIVOTS

Digital motions are always consistent with the proximal joint motions and with those of the wrist and hand or the ankle and foot, regardless of intermediate joint action. In the upper extremity, flexion with adduction of the fingers occurs with flexion of the wrist and shoulder adduction. Extension with abduction of the fingers occurs with wrist extension and shoulder abduction. The fingers rotate or glide toward the radial side consistently with radial motions of the wrist, supination, shoulder flexion, and external rotation. They rotate or glide toward the ulnar side with ulnar motions of the wrist, pronation, shoulder extension, and internal rotation.

Thumb flexion with adduction and external rotation of the first metacarpal occurs in the flexion–adduction–external rotation pattern. Thumb extension with adduction and external rotation of the first metacarpal occurs in the flexion–abduction–external rotation pattern. Thumb palmar abduction with abduction and internal rotation of the first metacarpal is combined with thumb extension in the extension – abduction – internal rotation pattern. Thumb opposition with abduction and internal rotation of the first metacarpal occurs in the extension-adduction-internal rotation pattern. Therefore, adduction of the thumb is consistent with external rotation and flexion of the shoulder; abduction of the thumb is consistent with internal rotation and extension of the shoulder; flexion of the thumb is consistent with adduction of the shoulder; and extension of the thumb is consistent with abduction of the shoulder.

In the lower extremity, extension with abduction

of the toes is combined with dorsiflexion of the foot and ankle and is consistent with flexion of the hip. Flexion with adduction of the toes is combined with plantar flexion and is consistent with hip extension. The toes rotate or glide toward the tibial side with inversion of the foot and hip adduction and external rotation. They rotate or glide toward the fibular side with eversion and hip abduction and internal rotation.

Rotation Patterns of Head and Neck and Upper Trunk

The rotation patterns of the head and neck and upper trunk are spiral in character. The major component is one of rotation from extreme left to extreme right or vice versa, thereby passing through a flexion phase and into an extension phase. Just as in the diagonal patterns, head and neck rotation is the key to upper trunk rotation. The motion is one of looking down and behind one shoulder then turning so as to look down and behind the opposite shoulder. The head and neck rotate as far as possible, and the trunk rotates from lateral hyperextension on one side to lateral hyperextension on the other side.

MAJOR MUSCLE COMPONENTS

The major muscle components of a given pattern are related by their topographical alignment upon the skeletal system and are primarily responsible for the movement. As an example, the flexion–adduction–external rotation pattern of the lower extremity consists primarily of muscles which are anteriorly and medially located. Specifically, the hip muscles are the iliopsoas group, gracilis, adductors longus and brevis, obturator externus, pectineus, and sartorius. When this pattern is performed with knee straight or with knee extension, the medial portion of the rectus femoris contributes its component of hip flexion. When knee extension is performed, the vastus medialis and the medial portion of the rectus femoris are primarily responsible. When the knee flexes, the medially located hamstrings, the semitendinosus, and the semimembranosus are primarily responsible. Distally, the anteriorly and medially located muscles are responsible for dorsiflexion with inversion of the ankle and foot and extension with abduction of the toes toward the tibial side. The muscles include the anterior tibial, extensor hallucis longus, extensor digitorum longus, abductor hallucis, extensor digitorum brevis, dorsal interossei, and lumbricales.

When the lower extremity is positioned with the hip in extension, abduction, and internal rotation, the knee straight, extended, or flexed, the ankle and foot plantar flexed in eversion and the toes flexed and adducted toward the fibular side, the topographical relationship of these muscles may be visualized. These are the major muscle components. Their action and cooperation is essential to performance of the described pattern.

The muscles secondarily responsible for a pattern are those most closely related by location and function. These muscles provide overlapping between patterns having one or two common components of action. The portions of muscles whose fibers are aligned with the muscles of a related pattern will contribute to that pattern although it may not be the optimal pattern for that particular muscle. For example, the extension–adduction–external rotation pattern is optimal for the gluteus maximus but it may contribute to the extension component of the extension–abduction–internal rotation pattern. The glutei medius and minimus are primarily responsible for this pattern but those fibers of the gluteus maximus which are aligned with those of the glutei medius and minimus will cooperate.

This type of overlapping is characteristic of the major muscle components of the proximal parts —the trunk, shoulder and hip. It contributes to the stability of these parts and is a sign of the versatility of muscles—that is, of their ability to contribute several components of action and their ability to contribute to several combinations of motion.

The versatility of muscles in patterns of facilitation is seen to progress from proximal to distal. Whereas, proximally a portion of a muscle may contribute to a related pattern, the muscles of the intermediate joints contribute specifically to two patterns related by two common components with reference to the proximal pivot. For example, the vastus medialis contributes to those patterns having components of adduction and external rotation thereby combining with hip flexion and hip extension. Overlapping of the proximal type occurs but to a lesser degree.

Distal muscles are more versatile in that they contribute specifically to two patterns, which are related by only one component with reference to the proximal joint. For example, the toe extensors contribute to both hip flexion patterns. The lumbricales are the most versatile of all muscles since they contribute to all patterns. This distal versatility contributes to dexterity and speed of movement as compared with proximal versatility which makes for stability.

LINE OF MOVEMENT

The spiral and diagonal patterns of facilitation provide for an optimal contraction of the major muscle components. A pattern of motion that is optimal for a specific "chain" of muscles allows these muscles to contract from their completely lengthened state to their completely shortened state, when the pattern is performed through the full range of motion. The optimal patterns for the individual muscles are shown in Tables 10 to 13, pp. 213–217.

In the starting position of a given pattern the major muscle components are in their completely lengthened state. The starting position is termed the lengthened range, the range of initiation, or the stretch range wherein the fibers of related muscles may be subjected to maximal stretch for facilitation. When the major muscle components contract the subject, or patient, moves the part from the lengthened range through the available range of motion to the shortened range. In the shortened range of the pattern, the major muscle components have reached their completely shortened state within the limits of the anatomical structure. The half-way or midpoint between the lengthened and shortened ranges is referred to as the middle range.

Positioning of a part in the lengthened range of a pattern requires consideration of all the components of motion from proximal to distal. The major components of flexion or extension are considered first. If a pattern has a component of flexion, the part is moved toward extension. Motion relative to the mid-line is next considered—if the pattern has a component of adduction, the part is moved toward abduction. Rotation is always considered last —if the pattern has a component of external rotation, the part is placed in internal rotation. In subjects who present less than normal range of passive motion, rotation should be considered first and last. Finally, positioning of the part is done smoothly with all three components considered and combined for diagonal placement.

As a pattern of motion is initiated, rotation enters the motion first as the spiral characteristic of the pattern, and the other two components combine to give the pattern a diagonal direction. A figurative cross may be drawn through the proximal joint or pivot to show the diagonal direction of the pattern. A cross having a perpendicular pole bisected at 90° by a horizontal pole should be rotated 45° either clockwise or counterclockwise to give the direction of the diagonals.

The diagonal line of movement is referred to as the "groove" of the pattern. It is the optimal line of movement produced by the optimal or maximal contraction of the major muscle components from their lengthened state to the shortened state. The normal subject readily demonstrates greater strength when he performs in the "groove" of the pattern than when the line of movement is to either side of the diagonal.

COOPERATIVE FUNCTION OF MUSCLES

Since three components of motion are considered with reference to all joints or pivots participating in a pattern, the major muscle components cooperatively contribute to the three components as far as their topographical location and structure will permit. The function of an individual muscle is a three-component action. The motion component which places the most stretch on a muscle determines its primary action component. The other motion components determine the secondary and tertiary action components. Thus, a muscle may be primarily a flexor, secondarily an adductor, and thirdly an external rotator.

Such a muscle is the psoas major, a major muscle component of the flexion–adduction–external rotation pattern of the lower extremity. Extension achieves the greatest amount of stretch, abduction the second greatest amount, and, finally, internal rotation completes the stretch. The psoas major has a primary action component of flexion, a secondary action component of adduction, and a tertiary action component of external rotation. When the pattern is performed from the lengthened range to the shortened range, the psoas major has contributed three components of action at the hip joint in cooperation with all major muscle components of the pattern.

A single muscle is not solely responsible for a single motion component. The individual muscle is augmented by other related muscles and, in turn, augments the action components of related muscles. The interrelationship of action components with reference to a specific pivot is finely shaded and graded and contributes to smoothness of motion. Following the example above, the psoas major is related topographically and functionally with the psoas minor and iliacus. This relationship is so close that they are commonly referred to as the iliopsoas group. These three muscles have common components of action in slightly varying degrees. The remaining muscle components of the pattern, the gracilis, adductors longus and brevis, pectineus, rectus femoris, sartorius, and obturator externus, all contribute flexion components even though minimal. These same muscles contribute adduction components in varying degrees and external rotation components in varying degrees. The obturator externus and sartorius are the primary intrinsic and extrinsic external rotators. The adductors longus and brevis, pectineus, and gracilis are the muscles having primary components of adduction. The gracilis contributes minimal external rotation. Deficiency of any muscle lessens the power with which the pattern is performed and disrupts the smoothness of the movement. A single motion component may be relatively weak while the other two are relatively strong depending on the primary action of the deficient major muscle components.

In the mature, normal subject, optimal contraction of the major muscle components occurs in sequence when the pattern is performed through the available range of motion. The normal timing is from distal to proximal. In the flexion–adduction–external rotation pattern of the lower extremity the motion is that of pulling the foot up and across the mid-line of the body as far as possible. This is true whether the knee remains straight, flexes, or extends. The range of motion occurring at the hip must necessarily vary in accordance with the knee motion used but the direction and goal of the pattern are the same. Smoothness of motion is dependent upon the foot moving first. If the foot moves last, the motion appears to be an afterthought. The contraction of muscles in sequence is related to coordination and is acquired in the developmental process.

Agonists and Antagonists

Optimal, sequential contractions of a "chain" of muscles implies true synergy of these muscles as they move the part through the available range of the pattern of motion. The pattern of muscles contracting toward their shortened state is termed the agonistic pattern. The pattern of muscles approaching their lengthened state in cooperation with those of the agonistic pattern is termed the antagonistic pattern.

The antagonistic pattern is composed of a "chain" of major muscle components having components of action which are exactly the opposite of those of the agonistic pattern. Their location is diagonally opposite those of the agonistic pattern. If a pattern is composed primarily of muscles which are anteriorly and medially located, the muscles of the antagonistic pattern are located posteriorly and

laterally. When flexion–adduction–external rotation of the lower extremity is considered the agonistic pattern, extension–abduction–internal rotation is the antagonistic pattern. Its major muscle components with reference to the hip pivot are the glutei medius and minimus. These muscles have action components of extension, abduction, and internal rotation and they are the direct antagonists of the iliopsoas group.

The lengthening reaction of the antagonistic pattern occurs from distal to proximal as demanded by the range of motion occurring in the agonistic pattern. As the shortened state of the agonistic pattern is reached, as the range of motion is completed, tension occurs in the antagonistic pattern as a range-limiting factor. This is most evident when the two-joint muscles are required to lengthen rather than to shorten. When the action of two-joint muscles is not considered, soft-tissue contact or ligamentous structures may become the range-limiting factor. The muscles of closely related patterns must also contribute lengthening reactions in line with the overlapping of muscle components. For example, in order for complete range of dorsiflexion and inversion of the foot and ankle to be achieved, the peroneus longus and peroneus brevis, among others, must lengthen. The first is primarily responsible to the extension–abduction–internal rotation pattern, while the latter is primarily responsible to the flexion–abduction–internal rotation pattern.

Summary of Muscle Function

In patterns of facilitation, the individual muscle contracts from its completely lengthened state to its completely shortened state in cooperation with the major muscle components of the pattern wherein it is located.

The individual muscle contributes three components of action as far as its topographical location and structure will allow.

The individual muscle lengthens completely in cooperation with its antagonists which are located diagonally opposite and have opposite components of action.

The individual muscle contributes to a related pattern as far as common components of motion and topographical location will allow. This contribution may be a shortening or lengthening reaction depending upon whether the related pattern is considered an agonist or an antagonist.

The individual muscle is not solely responsible for a single motion component of a pattern but is augmented by related muscles and, in turn, augments the action components of related muscles.

Deficiency of an individual muscle expresses itself to the greatest degree in relation to its primary action component, and to a lesser degree in relation to its secondary and tertiary components of action.

TYPES OF MUSCLE CONTRACTION

In techniques of proprioceptive neuromuscular facilitation, two types of muscle contraction are employed. An attempt by the subject to perform a pattern through any part of the range of motion is referred to as active motion with isotonic contraction of the muscles responsible for the motion. When a technique involves an attempt by the subject to hold a part still without permitting range of motion to occur, the muscular contraction is referred to as a "hold" or isometric contraction.

The specific techniques employ either isotonic contraction or isometric contraction and, in some instances, both types are used. Normal persons are capable of performing both types. Isotonic contractions are clearly related to movement; isometric contractions are clearly related to posture. In developing motor behavior the ability to move precedes the ability to maintain posture. Thus, isotonic contractions may be considered more primitive than isometric, "hold," contractions. In mature neuromuscular activity there must be a neat intermixture of both; movement is necessary to posture, and posture is necessary to movement.

INDICATIONS FOR PATTERNS

Patterns of facilitation are used as passive motion for determination of limitation of range of motion, as free active motion, guided active motion, or resisted motion, depending upon the indications for exercise. They may be performed through their full range of motion or in minimal parts of a range of motion as indicated. The goal of treatment is the coordinated performance of patterns of facilitation through a full range of motion and with a balance of power between antagonistic patterns of both diagonals of motion.

Patterns are described and illustrated for performance in the supine position. However, they may be performed, and usually should be, in any position which allows the desired range of motion to occur

with the greatest ease and strength. The normal subject is able to perform in a variety of positions. As he changes his position the relationship with gravity is altered accordingly. The influence and interaction of reflex mechanisms which underlie movement and posture are factors to be considered. For example, in the supine position the performance of mass extension of foot and ankle, knee, and hip (thrusting) is powerfully executed because gravity is said to "assist" the movement, and the position itself is reflexly favorable for extensor tonus. This same thrusting movement performed in the hands-knees position is far more difficult; now the subject must "overcome" gravity, and the position itself is reflexly favorable for flexion of the lower extremity. Thus, positioning for performance is a means of increasing or lessening the demand to be placed. Visual cues, too, influence performance and, in some instances, positioning to permit vision to lead or follow the movement may be of primary importance and may lessen the demand in performance.

Suggestions for Learning Patterns

1. Learn the components of motion by performing patterns as free, active motion in accordance with normal timing (see pattern description pages).
   a. Did rotation enter the motion first so that the motion was truly diagonal?
   b. Did the distal parts complete their full range of motion before the middle range was reached?
2. Proceed from one pattern to its directly antagonistic pattern.
3. Begin with the head and neck and upper trunk and proceed to upper extremities, lower trunk, and lower extremities.
4. Practice patterns in as many positions as possible—supine, sidelying (lateral), prone, hands-knees (creeping), sitting, kneeling, and standing.
5. Practice combining patterns as outlined in Tables 3–9.
6. Instruct and critcize other normal subjects in performance of patterns in free, active motion in accordance with normal timing.
7. Learn the major muscle components of each pattern with regard to each pivot of action or joint.

Individual patterns

ILLUSTRATIONS AND LEGENDS
(Figs. 1–37)

1. Each drawing portrays the full range of motion of a specific pattern. The initial position of the physical therapist, showing the lengthened range of the pattern, is in black. The middle range of the pattern is depicted by the dark gray figure and the shortened range of the pattern is shown by a light gray figure. The three positions show the motion characteristics of the pattern and the motion characteristics of the physical therapist. The physical therapist moves his body in order to allow the desired range of motion to occur.

2. The manual contacts portrayed in the illustration are optimum for the specific pattern. Certain variations are included, i.e., use of both hands distally, shifting to one hand proximally and one hand distally. The proximal hand may be shifted as needed during the patient's performance. Adaptations of manual contacts must be made when combinations of patterns are used during reinforcement.

3. The normal timing of the pattern is portrayed. The distal pivots of action have completed their range of motion by the time the middle range of the pattern is reached. Timing for emphasis, as a technique, will alter the range of motion of the proximal pivot when distal pivots are emphasized. The combination of motions and direction of the pattern remain unchanged. This variation in range is shown in Figures 36 and 37.

4. The spiral characteristic of the pattern is shown by the dotted lines placed medially or laterally on the extremities. The dotted lines on the trunk are intended to show the diagonal directions of the pattern with motion of the distal parts crossing the mid-line of the body.

5. The subject is portrayed in the supine position. Patterns may be performed in any position which permits the desired range of motion to occur.

6. Normal reinforcements by related patterns are not included in the illustrations. These are charted in Reference Tables 3–9 on pp. 209–212.

All information concerning a specific pattern is given for left or right extremities, or the trunk and neck motions to the left or right. For the specific pattern of the opposite side, left and right must be interchanged.

COMPONENTS OF MOTION

1. Description is from distal to proximal in accordance with normal timing.

2. Description, for the purpose of simplification, is limited to parts and major joints, rather than naming each individual joint which participates in the pattern of motion. Minor components of joint motion, such as gliding motions of metacarpal and carpal joints, contribute to the rotation and medialward of lateralward components of motion.

NORMAL TIMING

Action occurs from distal to proximal. Normal timing may be used with maximal resistance or may be performed as free active motion without contact with the subject by the physical therapist. See discussion of techniques of facilitation.

TIMING FOR EMPHASIS

Description is from proximal to distal in accordance with the normal process of development. Timing for emphasis alters range of motion of the proximal pivots when distal pivots are emphasized. This is illustrated in Figures 36 and 37. See discussion of techniques of facilitation.

COMMANDS

1. Preparatory commands must be varied in accordance with the age level and abilities of the

subject to cooperate. They may become superfluous after the subject has learned the desired patterns.

2. Action commands may be repeated as necessary in order to stimulate the patient to further effort. Sequence of commands used in applying other techniques must be learned in conjunction with those techniques.

While commands are described in terms of words and voice, other sensory cues are equally important, and, in many instances are more effective. Luring (the child) or urging the patient to look in the direction of the movement may be more meaningful than a dozen words of explanation or instruction. Quickly touching a part of the body will provide another clue to the patient. For example, a brisk tap on the upper left region of the chest will guide a patient as he performs neck flexion with rotation to the left. One externally applied stimulus may not be enough to facilitate response; two or three types of stimulus may produce far more response.

PATTERN ANALYSIS

1. The motion components and major muscle components are presented from proximal to distal in keeping with anatomical description.

2. The origins, insertions, and innervation of muscles is not included since this information is readily available to all.

3. The distal components of motion and major muscle components are listed only once for each pattern. They are presented with the patterns which require no motion of the intermediate joint or pivot.

4. Range-limiting factors are expressed in terms of the major muscle components of the antagonistic pattern. When the two-joint action of major muscles is not considered, soft tissue contact may become the range-limiting factor. Limitation by ligaments and joint structures is minimal unless hypermobility is present. Those ligaments which underlie the tendinous attachments of the major muscle components of the antagonistic pattern are potential range-limiting factors.

HEAD AND NECK

Flexion with Rotation to the Right

Fig. 1.

Antagonistic pattern

Extension with rotation to left (Fig. 2).

Components of motion

Head rotates toward right (axis on atlas), mandible depresses toward right, atlanto-occipital joint flexes toward right, and cervical spine flexes with rotation toward right so that chin approximates right clavicle.

Normal timing

Action occurs from distal to proximal, i.e., head rotates toward right (atlas on axis), mandible depresses as atlanto-occipital joint flexes toward right, and the cervical spine, which has been convex to right, flexes with rotation to right, and becomes convex to left.

Timing for emphasis

HEAD ROTATION TOWARD RIGHT

Allow beginning flexion to occur at atlanto-occipital joint with depression of mandible and beginning flexion of cervical spine to occur, but do not allow full range of cervical flexion with rotation toward right to occur, until head begins to rotate toward right.

Note: Resist stronger components of neck flexion.

DEPRESSION OF MANDIBLE AND FLEXION OF ATLANTO-OCCIPITAL JOINT

Allow beginning rotation of head and beginning flexion with rotation of cervical spine toward right to occur, but do not allow full range of head rotation and cervical flexion with rotation to occur, until mandible depresses with flexion occurring at atlanto-occipital joint.

Note: Resist stronger components of neck flexion, but guide weaker components through their optimal range of motion in accordance with normal timing.

CERVICAL FLEXION WITH ROTATION TOWARD RIGHT

Allow beginning head rotation and depression of mandible with flexion of atlanto-occipital joint to occur, but do not allow full range of head rotation and mandibular depression with atlanto-occipital flexion to occur, until cervical spine begins to flex with rotation toward right.

Note: Resist stronger components of neck flexion, but guide weaker components through their optimal range of motion in accordance with normal timing.

Manual contacts

RIGHT HAND

Pressure of medial palmar surface of hand and fingers under inferior surface of mandible on right between symphysis and right angle (Fig. 1).

LEFT HAND

Palmar surface of hand and fingers on left posterior-lateral aspect of skull so as to control rotation (Fig. 1).

Commands

PREPARATORY

"You are going to turn your head to the right, and pull it down and over toward the right, so that your chin touches your chest."

ACTION

"Turn your head!" "Pull your chin down!" "Pull your head down!"

Pattern analysis

HEAD ROTATION

Major muscle components: Right sternocleidomastoid, left rectus capitis lateralis, right rectus capitis anterior, right longus capitis—rotation component.

DEPRESSION OF MANDIBLE

Major muscle components: Right suprahyoid and infrahyoid muscles, platysma.

ATLANTO-OCCIPITAL JOINT FLEXION

Major muscle components: Right longus capitis—flexion component, right sternocleidomastoid.

Major muscle components: Right sternocleido-mastoid, longus capitis—flexion component, longus colli, scaleni—posterior, medius, anterior.

Note: The sternocleidomastoids are most versatile of all neck muscles and both muscles enter in when flexion is performed to right or left. When flexion is performed to right, right sternocleidomastoid contracts first, and as head approaches mid-line of body, left sternocleidomastoid contracts. Pattern of flexion to right must be first initiated by intrinsic muscles of pattern—if only sternocleidomastoids and supra- and infrahyoid muscles contract, motion is superficial, and shortened range of pattern will lack stability. Those supra- and infrahyoid muscles on left, the fibers of which are stretched by demanding neck flexion to right, will contribute to motion. Most closely related extremity pattern is extension—adduction–internal rotation of left upper extremity.

Range limiting factors

Tension or contracture of any of muscles of extension with rotation to left pattern (Fig. 2).

HEAD AND NECK

Extension with Rotation to the Left

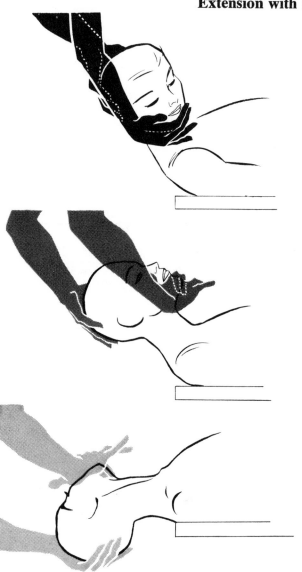

Fig. 2.

Antagonistic pattern

Flexion with rotation to right (Fig. 1).

Components of motion

Head rotates toward left, atlanto-occipital joint extends toward left, mandible elevates toward left, and cervical spine extends with rotation toward left, so that chin moves up and away from right clavicle.

Normal timing

Action occurs from distal to proximal, i.e., head rotates toward left, mandible elevates as atlanto-occipital joint extends, and cervical spine, which has been convex to left, extends with rotation to left, and becomes convex to right.

Timing for emphasis

HEAD ROTATION TOWARD LEFT

Allow beginning extension to occur at atlanto-occipital joint with elevation of mandible, but do not allow full range of cervical extension with rotation toward left to occur, until head begins to rotate toward left.

Note: Resist stronger components of neck extension.

ELEVATION OF MANDIBLE AND EXTENSION OF ATLANTO-OCCIPITAL JOINT

Allow beginning rotation of head and beginning extension with rotation of cervical spine to occur toward left, but do not allow full range of head rotation and cervical extension with rotation to occur, until mandible elevates with extension occurring at atlanto-occipital joint.

Note: Resist stronger components of neck extension, but guide weaker components through their optimal range of motion in accordance with normal timing.

CERVICAL EXTENSION WITH ROTATION TOWARD LEFT

Allow beginning head rotation and elevation of mandible with extension of atlanto-occipital joint to occur, but do not allow full range of head rotation and mandibular elevation with atlanto-occipital extension to occur, until cervical spine begins to extend with rotation toward left.

Note: Resist stronger distal components of pattern but guide weaker components through their optimal range of motion in accordance with normal timing.

Manual contacts

RIGHT HAND

Pressure of lateral palmar surface of hand and fingers on superior surface of mandible on left between symphysis and left angle (Fig. 2).

LEFT HAND

Pressure of palmar surface of hand and fingers on left posterior-lateral surface of the occiput and cervical region (Fig. 2).

Commands

PREPARATORY

"You are going to turn your head to the left, and lift your chin up and away from your chest."

ACTION

"Turn your head." "Lift your chin up!" "Push your head back!"

Pattern analysis

HEAD ROTATION

Major muscle components: Left-obliquus capitis superior, obliquus capitis inferior—rotation component, splenius capitis, longissimus capitis, semispinalis capitis, trapezius—upper portion.

ELEVATION OF MANDIBLE AND ATLANTO-OCCIPITAL JOINT EXTENSION

Major muscle components: Left obliquus capitis inferior—extension component, rectus capitis posterior major, rectus capitis posterior minor, semispinalis capitus, longissimus capitis, splenius capitis.

CERVICAL EXTENSION WITH ROTATION

Major muscle components: Left semisplinalis capitis, longissimus capitis, longissimus cervicis, iliocostalis cervicis, splenius capitis, splenius cervicis, interspinales, intertransversarii, trapezius—upper portion. Right semispinalis cervicis, multifidus.

Note: Since rotation of the vertebral column becomes more evident when motion of entire column is considered, the rotation characteristic of components of muscle action becomes apparent when the pattern is performed through full range. Major cervical muscles contribute a component of rotation as well as a component of flexion or extension. Also, because of overlapping of origins and insertions, there is overlapping of components in contributing muscle groups. More laterally located extensor muscles have a stronger component of rotation, but intrinsic rotation is the responsibility of specific rotator muscles such as obliquus capitis inferior and multifidus. Overlapping of action between muscles of left and muscles of right is apparent in flexion and extension patterns and is also characteristic of upper trunk patterns. The most closely related extremity pattern is flexion–abduction–external rotation of left upper extremity.

Range limiting factors

Tension or contracture of any of muscles of flexion with rotation to right pattern (Fig. 1).

HEAD AND NECK

Rotation to the Right

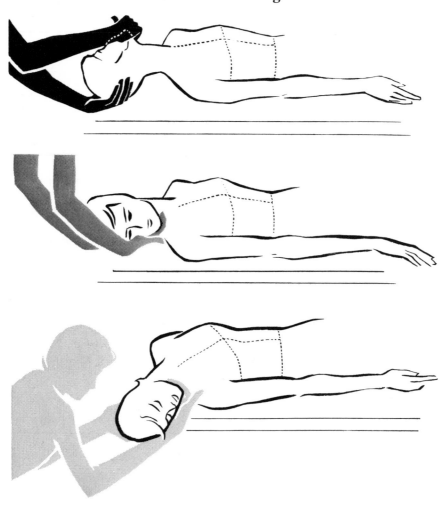

Fig. 3.

Antagonistic pattern

Rotation to left (motion components, major muscle components, and manual contacts are exactly opposite).

Components of motion

Head rotates toward right, mandible depresses and rotates from left to right, atlanto-occipital joint flexes toward right, and cervical spine rotates through flexion and into extension toward right. Cervical spine, which has been convex to right, rotates and becomes convex to left.

Normal timing

Action occurs from distal to proximal, i.e., head rotates toward right, as atlanto-occipital joint flexes and right mandible depresses and approaches right shoulder, as cervical spine rotates and becomes convex to left.

Timing for emphasis

HEAD ROTATION TOWARD RIGHT

Allow beginning flexion of atlanto-occipital joint and mandibular depression with rotation of cervical spine, but do not allow full range of mandibular depression and cervical rotation to occur, until head begins to rotate.

Note: Resist stronger components of neck rotation.

DEPRESSION OF MANDIBLE AND FLEXION OF ATLANTO-OCCIPITAL JOINT

Allow beginning head rotation and cervical spine rotation through flexion to occur, but do not allow full range of head rotation and cervical spine rotation to occur, until mandible begins to depress with flexion of atlanto-occipital joint.

Note: Resist stronger components of neck rotation, but guide weaker components through their optimal range of motion in accordance with normal timing.

CERVICAL SPINE ROTATION (THROUGH FLEXION INTO EXTENSION)

Allow beginning head rotation and mandibular depression with atlanto-occipital flexion to occur toward the right, but do not allow full range of head rotation and mandibular depression with atlanto-occipital flexion to occur, until cervical spine begins to rotate through flexion and toward extension to right.

Note: Resist stronger components of head and neck rotation, but guide weaker components through their optimal range of motion in accordance with normal timing.

Manual contacts

RIGHT HAND

Pressure of medial palmar surface of hand and fingers under inferior border of mandible on right; tips of fingers near symphysis, so as to control flexion and rotation components (Fig. 3).

LEFT HAND

Pressure of lateral palmar surface of hand and fingers on right posterior-lateral surface of skull, between mastoid process and occiput, and extending downward to lateral cervical extensors (Fig. 3).

Commands

PREPARATORY

"You are going to turn your head, so that your chin is touching your right shoulder, as if you were going to look down and behind your shoulder."

ACTION

"Turn it!" "Get your chin on your shoulder!" "Push your head back!"

Pattern analysis

HEAD ROTATION

Major muscle components: Right rectus capitis anterior, left rectus capitis lateralis, right sternocleidomastoid.

DEPRESSION OF MANDIBLE AND FLEXION OF ATLANTO-OCCIPITAL JOINT

Major muscle components: More lateral right supra- and infrahyoid muscles and right sternocleidomastoid.

CERVICAL SPINE ROTATION (THROUGH FLEXION INTO EXTENSION)

Major muscle components: Right scaleni medius and posterior, longissimus capitis, longissimus cervicis, iliocostalis cervicis, splenius capitis, splenius cervicis, and semispinalis capitis.

Note: Both sternocleidomastoids act in rotation of neck to left or to right. When neck is rotated from left to right, right sternocleidomastoid acts first, and as head passes the mid-line of body, left sternocleidomastoid contracts and maintains head in shortened range of rotation to right. Rotation cannot be separated from major components of flexion or extension of vertebral column. In rotation pattern, first components are related to flexion with rotation in intiation of pattern, shortened range of pattern requires components of extension with rotation. More lateral muscles contribute most strongly to rotation. Rotation pattern is optimum for rotation components of contributing muscles, but is not optimum for flexion or extension components. Most closely related extremity patterns are extension–abduction–internal rotation of right upper extremity and flexion–adduction–external rotation of left upper extremity.

Range limiting factors

Tension or contracture of any muscles of rotation to left pattern and of flexion to left pattern and neck extension to left pattern.

UPPER TRUNK (Superior Region)

Flexion with Rotation to the Right

Fig. 4.

Antagonistic pattern

Upper trunk extension with rotation to the left (Fig. 5).

Components of motion

Head rotates toward right, atlanto-occipital joint flexes with mandible depressing toward right, cervical and dorsal regions of spine, which have been convex to right, flex with rotation and become convex to left. Forehead approaches right hip.

Normal timing

Action occurs from distal to proximal, i.e., head rotation, then atlanto-occipital flexion with mandibular depression, then cervical spine flexion with rotation, and dorsal spine flexion with rotation.

Timing for emphasis

DORSAL SPINE FLEXION WITH ROTATION

Allow beginning contraction of components of neck flexion with rotation to right pattern, in accordance with normal timing, but do not allow full range of these components to occur, until abdominals contract and dorsal spine begins to flex and rotate toward right.

Note: Resist stronger components of neck flexion, but guide weaker components through their optimal range of motion in accordance with normal timing.

Manual contacts

LEFT HAND

Pressure of palmar surface of hand and fingers on right anterior-lateral aspect of patient's forehead (Fig. 4).

RIGHT HAND

Palmar surface of hand and fingers cupped over dorsal-ulnar aspect of fingers and wrist of patient's right hand (Fig. 4).

Commands

PREPARATORY

"You are going to turn your head, and pull yourself up and over toward your right hip."

ACTION

"Pull up and over!" "Turn it!" "Pull your chin down!" "Pull your head down!" "Pull your arms toward your right hip!"

Pattern analysis

Motion components and major muscle components of neck flexion with rotation to the right (Fig. 1).

Major muscle components: Left external oblique, right internal oblique, rectus abdominis—right portion, left transversus thoraciis, right internal intercostals, right subcostals (quadratus lumborum).

Note: Most closely related extremity patterns are extension–adduction–internal rotation of left upper extremity, extension–abduction–internal rotation of right upper extremity, flexion–adduction–external rotation of right lower extremity and flexion–abduction–internal rotation of left lower extremity. Lower trunk pattern which is most closely related is lower trunk flexion with rotation to left. When this pattern is combined with upper trunk flexion with rotation to right, crossing of rotation components occurs at dorsolumbar junction of vertebral column. When upper trunk flexion with rotation to right is combined with lower trunk flexion with rotation to right, all the oblique abdominal muscles contract.

Range limiting factors

Tension or contracture of any muscles of upper trunk extension with rotation to left pattern (Fig. 5).

Illustration note

Figure 4 portrays reinforcement of upper trunk flexion with rotation to right by combined upper extremity patterns which are most closely related—those of extension–adduction–internal rotation of left upper extremity and extension–abduction–internal rotation of right upper extremity. Extension–adduction–internal rotation pattern contributes to flexion component of upper trunk while extension–abduction–internal rotation pattern contributes to rotation of upper trunk. These combined patterns are referred to as "chopping."

If pattern is to be performed without resistance to "chopping," manual contacts for head and neck flexion with rotation to right pattern may be used.

UPPER TRUNK (Superior Region)

Extension with Rotation to the Left

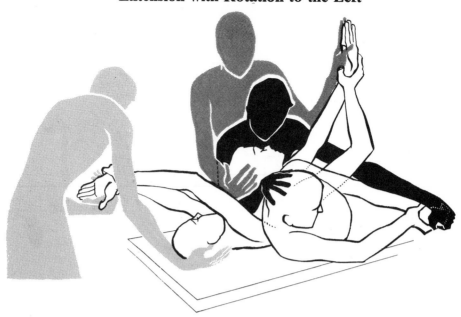

Fig. 5.

Antagonistic pattern

Upper trunk flexion with rotation to the right (Fig. 4).

Components of motion

Head rotates toward left, atlanto-occipital joint extends toward left with mandible elevating toward left, cervical and dorsal regions of spine, which have been convex to left, extend with rotation and become convex to right. Forehead moves away from right hip.

Normal timing

Action occurs from distal to proximal, i.e., head rotation, then atlanto-occipital extension with mandibular elevation, then cervical spine extension with rotation, and dorsal spine extension with rotation.

Timing for emphasis

DORSAL SPINE EXTENSION WITH ROTATION

Allow beginning contraction of all components of neck extension with rotation to left, in accordance with normal timing, but do not allow full range of their components to occur, until dorsal extensors on left contract and dorsal spine begins to extend and rotate toward left.

Note: Resist components of neck extension, but guide weaker components through their optimal range of motion in accordance with normal timing.

Manual contacts

LEFT HAND

Palmar surface of hand and fingers cupped over dorsal-radial aspect of fingers and wrist of patient's left hand (Fig. 5).

RIGHT HAND

Pressure of palmar surface of hand and fingers on left posterior-lateral aspect of patient's head (Fig. 5).

Commands

PREPARATORY

"You are going to turn your head, and push it up and away from your right hip, so that you are looking up and over your left shoulder."

ACTION

"Push up and over!" "Turn your head!" "Lift your arms up!" "Push your head away and up!" "Straighten your back!"

Pattern analysis

Motion components and major muscle components of neck extension with rotation to left (Fig. 2).

DORSAL SPINE EXTENSION WITH ROTATION

Major muscle components: Left spinalis dorsi, longissimus dorsi, iliocostalis dorsi, iliocostalis lumborum, quadratus lumborum, interspinales, intertransversarii, serratus posterior superior, external intercostals. Right semispinalis dorsi, levatores costarum, multifidus, rotatores, serratus posterior inferior. Transversus abdominis.

Note: Most closely related patterns are flexion–abduction–external rotation of left upper extremity, flexion–adduction–external rotation of right upper extremity, extension–abduction–internal rotation of left lower extremity and extension–adduction–external rotation of right lower extremity. When lower trunk extension with rotation to left is combined with upper trunk extension to left, entire spine extends with rotation toward left with right convexity occurring in shortened range of combined patterns. When lower trunk extension with rotation to right is combined with upper trunk extension with rotation to left, dorsal spine becomes convex to right and lumbar spine becomes convex to left with crossing of rotation occurring at dorso-lumbar junction.

Illustration note

Illustration portrays reinforcement of upper trunk extension with rotation to left by combined upper extremity patterns which are most closely related—those of flexion–abduction–external rotation of left upper extremity and flexion–adduction–external rotation of right upper extremity. Flexion–abduction–external rotation pattern contributes to component of extension of upper trunk while flexion–adduction–external rotation pattern contributes to rotation of upper trunk. These combined upper extremity patterns are referred to as "lifting."

If pattern is to be performed without resistance to "lifting," manual contacts for head and neck extension with rotation to left pattern may be used.

Range limiting factors

Tension or contracture of any muscles of upper trunk flexion with rotation to right pattern (Fig. 4).

UPPER TRUNK (Superior Region)

Rotation to the Right

Antagonistic pattern

Upper trunk rotation to the left. (Motion components, major muscle components, and manual contacts are exactly opposite.)

Components of motion

Head rotates toward right, atlanto-occipital joint flexes with depression and rotation of the mandible from left to right, cervical and dorsal spine rotates through flexion and into extension toward right. Cervical and dorsal regions of spine, which have been convex to right, rotate and become convex to left.

Normal timing

Action occurs from distal to proximal, i.e., head rotation, atlanto-occipital flexion with mandibular depression and rotation, then cervical and dorsal spine rotation through flexion into extension.

Timing for emphasis

UPPER TRUNK ROTATION TO RIGHT

Allow beginning contraction of all components of head and neck rotation to right, in accordance with normal timing, but do not allow full range to occur, until contraction proceeds in the abdominals and laterally into extensors.

Note: Resist stronger components of rotation, but guide weaker components through their optimal range of motion in accordance with normal timing.

Manual contacts

Left and right hands as in head and neck rotation to the right (Fig. 3).

Commands

PREPARATORY

"You are going to turn your head to the right, and twist your body, so as to look down and behind your right shoulder."

ACTION

"Turn!" "Pull your chin to your shoulder!" "Pull your chin down!" "Push your head back!"

Pattern analysis

Motion components and major muscle components of head and neck rotation to right.

UPPER TRUNK ROTATION TO RIGHT

Major muscle components: Rotation components of trunk flexors—left external oblique, right internal oblique, transversus abdominus. Rotation component of trunk extensors—right iliocostalis dorsi, iliocostalis lumborum, quadratus lumborum.

Note: Upper extremity patterns most closely related to upper trunk rotation to right are extension–abduction–internal rotation of right upper extremity and flexion–adduction–external rotation of left upper extremity. These are the patterns which contribute to rotation of upper trunk flexion to right pattern and upper trunk extension to right pattern. Since upper trunk rotation to right is primarily a pattern of rotation but has components of flexion and extension, these upper extremity patterns combine most effectively to reinforce rotation of upper trunk. Pattern may be performed with manual contacts applied to head and one upper extremity, or to both upper extremities with free active motion of head and neck. This pattern is optimum for quadratus lumborum on right. Extension–abduction–internal rotation of right lower extremity is most closely related.

Range limiting factors

Tension or contracture of any muscles of rotation to left pattern.

LOWER TRUNK (Inferior Region)

Flexion with Rotation to the Left

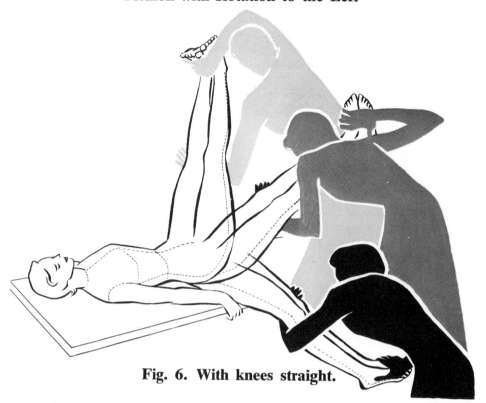

Fig. 6. With knees straight.

Antagonistic pattern

Lower trunk extension with rotation to right (Figs. 9–11).

Components of motion

Lower extremities, in close approximation, flex and rotate to left, requiring flexion–abduction–internal rotation of left lower extremity and flexion–adduction–external rotation of right lower extremity including all of their respective components of motion. Intermediate joints, knees, may remain straight, may flex or extend. Pelvis rotates, brim moving upward and to left. Lumbar spine, which has been convex to left, flexes with rotation and becomes convex to right.

Normal timing

Action occurs from distal to proximal, i.e., action proceeds from distal to proximal in relation to lower extremities, toes, feet, ankles, knees (if intermediate joint motion is desired), then hips, pelvic rotation, and lumbar flexion with rotation to left.

Timing for emphasis

Allow beginning rotation to occur at toes, ankles, knees, and hips, but do not allow full range of lower extremity components to occur, until pelvis begins to rotate toward left and lumbar spine begins to flex with rotation toward left.
Note: When intermediate joint action is used, that is knee flexion or extension, pelvic rotation and lumbar

flexion is delayed because of time required for action of intermediate joints.

Manual contacts

RIGHT HAND

Pressure of palmar surface of hand and fingers on dorsal aspect of both feet, especially left foot. Internal malleoli should be in close approximation (Fig. 8). If patient has no active motion below ankles, right hand may be used to grip both heels so as to control rotation at hips.

LEFT HAND

Pressure of palmar surface of hand, fingers, and forearm on anterior surface of both thighs proximal to knee joint. Patient's knees should be in close approximation. If patient has difficulty in initiating range of hip motions, left hand may be placed on posterior aspect of thighs proximal to popliteal spaces (Fig. 8).

Commands

PREPARATORY

"You are going to turn your heels away from me, and pull your feet up and across your body." "Keep your knees straight," or "Bend your knees," or "Straighten your knees."

ACTION

"Pull!" "Pull your feet up!" "Keep your knees straight!" ("Bend your knees!" or "Straighten your knees!") "Pull them up and away from me!"

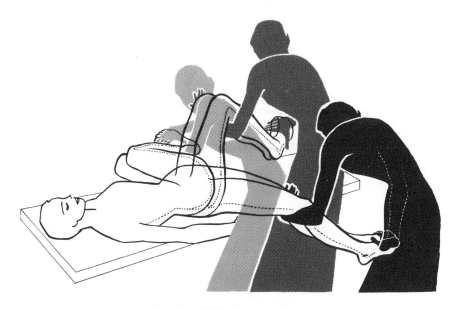

Fig. 7. With knees flexing.

Pattern analysis

LOWER EXTREMITIES

Motion components and major muscle components: Same as flexion–abduction–internal rotation of left (Figs. 30–32), and flexion–adduction–external rotation of right (Figs. 24–26).

PELVIC ROTATION AND LUMBAR SPINE FLEXION WITH ROTATION TO LEFT

Major muscle components: Left external oblique, rectus abdominis–left portion, quadratus lumborum. Right internal oblique.

Note: Most closely related upper extremity patterns are extension–adduction–internal rotation of left and flexion–adduction–external rotation of right. Extension–adduction–internal rotation of left contributes to flexion component, flexion–adduction–external rotation of right contributes to rotation component. Upper trunk flexion with rotation to right requires action of same oblique abdominal muscles and upper trunk flexion with rotation to left requires action of all abdominal muscles. Neck flexion with rotation to right and to left are related in the same manner as the upper trunk flexion patterns.

Range limiting factors

Tension or contracture of any muscles of lower trunk extension with rotation to right pattern (Figs. 9–11).

Fig. 8. With knees extending.

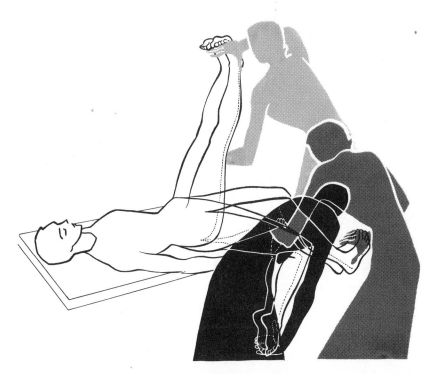

LOWER TRUNK (Inferior Region)
Extension with Rotation to the Right

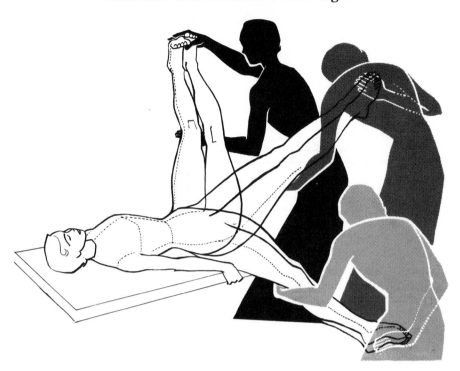

Fig. 9. With knees straight.

Antagonistic pattern

Lower trunk flexion with rotation to the left. (Figs. 6–8).

Components of motion

Lower extremities, in close approximation, extend and rotate to right, requiring extension–abduction–internal rotation of right lower extremity and extension–adduction–external roation of left lower extremity, including all of their respective components of motion. Intermediate joints may remain straight, flex, or extend. Pelvis rotates, brim moving downward and to right. Lumbar spine, which has been convex to right, extends with rotation and becomes convex to left.

Normal timing

Action occurs from distal to proximal, i.e., action proceeds from distal to proximal in relation to lower extremities, toes, feet, ankles, knees (if intermediate joint motion is desired), then hips, pelvic rotation, and lumbar extension with rotation to right.

Timing for emphasis

Allow beginning rotation to occur at toes, ankles, knees, and hips, but do not allow full range of lower extremity components to occur, until pelvis rotates toward right and lumbar spine begins to extend with rotation toward right.

Note: When intermediate joint action is used, that is, knee flexion or knee extension, pelvic rotation and lumbar spine extension is delayed because of time required for action of intermediate joints. Resist stronger distal components, but guide weaker distal components through their optimal range of motion in accordance with normal timing.

Manual contacts

RIGHT HAND

Pressure of palmar surface of hand and fingers on plantar surface of both feet, especially right foot. Internal malleoli should be in close approximation (Fig. 11). If patient has no active motion below ankle, right hand may be used to grip both heels so as to control rotation at hips.

LEFT HAND

Pressure of palmar surface of hand, fingers and forearm on posterior surface of both thighs proximal to popliteal space. Patient's knees should be in close approximation (Fig. 11).

Commands

PREPARATORY

"You are going to turn your heels toward me, and push your feet down and over toward me." "Keep your knees straight," or "Bend your knees," or "Straighten your knees."

Fig. 10. With knees extending.

"Push!" "Turn your heels!" "Push your feet down!" "Push toward me!" "Keep your knees straight!" ("Bend your knees!" or "Straighten your knees!")

Pattern analysis

LOWER EXTREMITIES

Motion components and major muscle components: Same as extension–abduction–internal rotation pattern of the right (Figs. 27–29) and extension–adduction–external rotation of left (Figs. 33–35).

PELVIC ROTATION AND LUMBAR SPINE
EXTENSION WITH ROTATION TO RIGHT

Major muscle components: Right sacrospinalis, iliocostalis lumborum, quadratus lumborum, interspinales, intertransversarii, longissimus dorsi, spinalis dorsi; left multifidus, rotatores. Right dorsal extensors and rotators will enter motion if pattern is initiated from range sufficient to demand their response.

Note: Upper trunk extension with rotation to right is most closely related upper trunk pattern; neck extension with rotation to right is most closely related neck pattern. Most closely related upper extremity patterns are flexion–abduction–external rotation of left upper extremity which reinforces extension component, and extension–abduction–internal rotation of right upper extremity which reinforces rotation component.

Range limiting factor

Tension or contracture of any muscles of lower trunk flexion with rotation to left pattern (Figs. 6–8).

Fig. 11. With knees flexing.

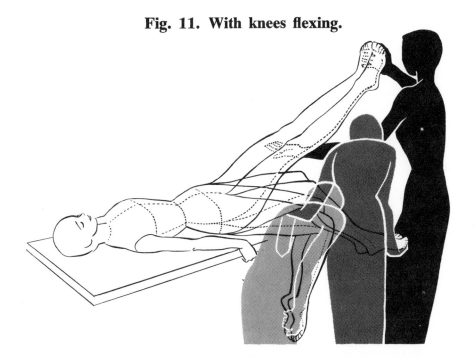

UPPER EXTREMITY
Flexion–Adduction–External Rotation

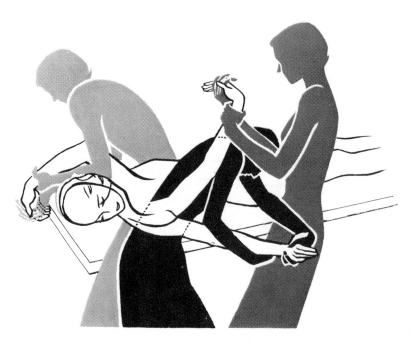

Fig. 12. With elbow straight.

Antagonistic pattern

Extension–abduction–internal rotation (with elbow straight) (Fig. 15).

Components of motion

Fingers flex and adduct toward radial side (lateral fingers more than medial), thumb externally rotates, flexes, and adducts toward radial side, wrist supinates and flexes toward radial side, forearm supinates, elbow remains straight, shoulder flexes, adducts and externally rotates with scapula rotating, abducting (inferior angle), and elevating anteriorly (acromion) and clavicle approximates sternum with rotation and elevation anteriorly.

Normal timing

Action is from distal to proximal, i.e., action occurs first at fingers, thumb, wrist, and forearm, then at shoulder, scapula, and clavicle.

Timing for emphasis

SCAPULA AND CLAVICLE

Allow beginning rotation to occur at fingers, wrist, forearm, and shoulder, but do not allow full range of finger flexion with adduction toward radial side, wrist flexion toward radial side, forearm supination, and shoulder flexion–adduction–external rotation to occur, until scapula begins to rotate, abduct and elevate anteriorly.

Note: If normal timing is prevented by excessive resistance to weak distal components, action cannot occur proximally. Resist stronger distal components, but guide weaker distal components through their optimal range of motion in accordance with normal timing.

SHOULDER

Allow beginning rotation to occur at fingers, wrist, forearm, and shoulder, but do not allow full range of finger flexion with adduction toward radial side, wrist flexion toward radial side, and forearm supination to occur, until shoulder begins to flex and adduct in external rotation.

Note: Resist stronger proximal and distal components, but guide weaker distal components through their optimal range of motion in accordance with normal timing.

FOREARM

Allow beginning rotation to occur at fingers, thumb, wrist, forearm, and shoulder, but do not allow full range of finger flexion with adduction toward radial side, wrist flexion toward radial side, and shoulder flexion–adduction–external rotation to occur, until forearm begins to supinate.

Note: Resist stronger proximal and distal components, but guide weaker distal components through their optimal range of motion in accordance with normal timing.

Patterns of facilitation

WRIST

Allow beginning rotation to occur at fingers, thumb, wrist, forearm, and shoulder, but do not allow full range of finger flexion toward radial side, forearm supination, and shoulder flexion–adduction–external rotation to occur, until wrist begins to flex toward radial side.

Note: Resist stronger proximal and distal components, but guide weaker distal components through their optimal range of motion in accordance with normal timing.

FINGERS

Allow beginning rotation to occur at fingers, thumb, wrist, forearm, and shoulder, but do not allow full range of wrist flexion toward radial side, forearm supination, and shoulder flexion–adduction–external rotation to occur, until fingers begin to flex and adduct toward radial side.

Note: Resist stronger proximal components. Emphasis may be placed on metacarpal-phalangeal joints, or interphalangeal joints, or emphasis may be placed upon a specific joint of an individual digit.

THUMB

Allow beginning rotation to occur at fingers, wrist, forearm, and shoulder, but do not allow full range of other components to occur, until thumb begins to flex and adduct. Components of fingers and wrist must be allowed to move through range after motion is initiated at thumb. In shortened range of pattern thumb is flexed, adducted and externally rotated toward second metacarpal.

Note: Resist stronger finger, wrist, and proximal components.

Manual contacts
LEFT HAND

Placed in palm of patient's right hand, so that patient may grasp with fingers and thumb, and so that wrist may flex toward radial side (Fig. 12).

RIGHT HAND

For emphasis of distal joints: Grip with pressure of palmar surface over distal anterior aspect of forearm, so as to control supination and proximal components of motion (Fig. 12).

For emphasis of shoulder: Pressure of palmar surface on anterior-medial surface of patient's arm, so as to control external rotation and proximal components of motion.

For emphasis of scapula: Pressure of palmar surface of hand over anterior aspect of patient's shoulder proximal to acromion process.

For mass closing of hand and for emphasis of thumb motion: Grasp right thumb of patient with thumb and index finger of left hand, contact of thumb and finger medially and laterally at interphalangeal

joint of thumb. Place right fingers and hand on palmar surface of fingers of patient's right hand. Flexion of all fingers and thumb flexion–adduction may be resisted. Physical therapist's right hand prevents range of motion occurring proximally until fingers flex and adduct.

Commands
PREPARATORY

"You are to squeeze my hand, turn it, and pull my hand up and across your face, keeping your elbow straight."

ACTION

"Pull!" "Squeeze my hand!" "Turn it!" "Pull up across your face!" "Keep your elbow straight!"

Pattern analysis
SCAPULA

Motion components: Rotation, abduction (inferior angle), elevation anteriorly (acromion).
Major muscle components: Serratus anterior.

SHOULDER

Motion components: Flexion, adduction, external rotation.
Major muscle components: Pectoralis major—clavicular portion, deltoid—anterior portion, coracobrachialis, biceps brachii—shoulder flexion component.

FOREARM

Motion components: Supination.
Major muscle components: Supinator.

WRIST

Motion components: Flexion toward radial side.
Major muscle components: Flexor carpi radialis, palmaris longus.

FINGERS

Motion components: Flexion, adduction toward radial side.
Major muscle components: Flexor digitorum superficialis, flexor digitorum profundus, flexor digiti quinti brevis, opponens digiti quinti, palmar interossei, lumbricales.

THUMB

Motion components: Flexion, adduction with rotation toward second metacarpal.
Major muscle components: Flexor pollicis longus, flexor pollicis brevis, adductores pollicis

Range limiting factor

Tension or contracture of any muscles of the extension–abduction–internal rotation (with elbow straight) pattern (Fig. 15).

UPPER EXTREMITY
Flexion–Adduction–External Rotation

Fig. 13. With elbow flexion.

Antagonistic pattern

Extension–abduction–internal rotation (with elbow extension) (Fig. 16).

Components of motion

Fingers flex and adduct toward radial side (lateral fingers more than medial), thumb externally rotates, flexes, and adducts toward radial side, wrist supinates and flexes toward radial side, forearm supinates, elbow flexes, shoulder flexes, adducts, and externally rotates with scapula rotating, abducting (inferior angle), and elevating anteriorly (acromion), and clavicle approximates sternum with rotation and elevation anteriorly.

Normal timing

Action is from distal to proximal, i.e., action occurs first at fingers, thumb, wrist, forearm, and elbow, then at shoulder, scapula, and clavicle.

Timing for emphasis
SCAPULA AND CLAVICLE

Allow beginning rotation to occur at fingers, thumb, wrist, forearm, elbow, and shoulder, but do not allow full range of finger flexion with adduction toward radial side, wrist flexion toward radial side, forearm supination, elbow flexion, and shoulder flexion–adduction to occur, until scapula begins to rotate, abduct, and elevate anteriorly.

Note: If normal timing is prevented by excessive resistance to weak distal components, action cannot occur proximally. Resist stronger distal components, but guide weaker distal components through their optimal range of motion in accordance with normal timing.

SHOULDER

Allow bgeinning rotation to occur at fingers, thumb, wrist, forearm, elbow, and shoulder, but do not allow full range of finger flexion with adduction toward radial side, wrist flexion toward radial side, forearm supination, elbow flexion and scapular rotation to occur, until shoulder begins to flex and adduct in external rotation.

Note: Resist stronger proximal and distal components, but guide weaker distal components through their optimal range of motion in accordance with normal timing.

ELBOW

Allow beginning rotation to occur at fingers, thumb, wrist, forearm, elbow, and shoulder, but do not allow full range of finger flexion with adduction toward radial side, wrist flexion toward radial side, forearm supination, shoulder flexion–adduction and scapular rotation to occur, until elbow begins to flex.

Note: Resist stronger distal and proximal components, but guide weaker distal components through their optimal range of motion in accordance with normal timing.

FOREARM

Allow beginning rotation to occur at fingers, thumb, wrist, forearm, elbow, and shoulder, but do not allow full range of finger flexion with adduction toward radial side, wrist flexion toward radial side, elbow flexion, shoulder flexion–adduction with scapular rotation to occur, until forearm begins to supinate.

Note: Resist stronger proximal and distal components, but guide weaker components through their optimal range of motion in accordance with normal timing.

WRIST

Allow beginning rotation to occur at fingers, thumb, wrist, forearm, elbow, and shoulder, but do not allow full range of finger flexion with adduction toward radial side, forearm supination, elbow flexion, shoulder flexion–adduction with scapular rotation to occur, until wrist begins to flex toward radial side.

Note: Resist stronger distal and proximal components, but guide weaker distal components through their optimal range of motion in accordance with normal timing.

FINGERS

Allow beginning rotation to occur at fingers, thumb, wrist, forearm, elbow, and shoulder, but do not allow full range of wrist flexion toward radial side, forearm supination, elbow flexion, and shoulder flexion–adduction to occur, until fingers begin to flex and adduct toward radial side.

Note: Resist stronger proximal components. Emphasis may be placed upon metacarpal-phalangeal joints, or interphalangeal joints, or emphasis may be placed upon a specific joint of an individual digit.

THUMB

Allow beginning rotation to occur at fingers, thumb, wrist, forearm, elbow, and shoulder, but do not allow full range of finger flexion with adduction toward radial side, wrist flexion toward radial side, forearm supination, elbow flexion, and shoulder flexion–adduction to occur, until thumb begins to flex and adduct. Components of fingers and wrist must be allowed to move through range after motion is initiated at thumb. In shortened range of pattern thumb is flexed, adducted, and externally rotated toward second metacarpal.

Note: Resist stronger finger, wrist, and proximal components.

Manual contacts

LEFT HAND

Placed in palm of patient's right hand, so that patient may grasp with fingers and thumb, and so that wrist may flex toward radial side (Fig. 13).

RIGHT HAND

For emphasis of distal joints: Grip with pressure of palmar surface over distal anterior aspect of fore-arm, so as to control supination and proximal components of motion.

For emphasis of elbow: Pressure of palmar surface of hand on anterior-medial surface of arm, so as to control external rotation and proximal components of motion (Fig. 13).

For emphasis of shoulder: Same as for elbow emphasis.

For emphasis of scapula: Pressure of palmar surface of hand over anterior aspect of shoulder proximal to acromion process.

Mass closing of hand and for emphasis of thumb motion: Grasp right thumb of patient with thumb and index finger of left hand. Place right fingers and hand on palmar surface of fingers of patient's right hand. Flexion of all fingers and thumb flexion–adduction may be resisted. Physical therapist's right hand prevents range of motion occurring proximally until fingers flex and adduct.

Commands

PREPARATORY

"You are to squeeze my hand, turn it, and bend your elbow, and pull my hand up and across your face."

ACTION

"Pull!" "Squeeze my hand!" "Turn it!" "Bend your elbow!" "Pull up across your face!"

Pattern analysis

SCAPULA

Motion components: Rotation, abduction (inferior angle), elevation anteriorly (acromion).

Major muscle components: Serratus anterior.

SHOULDER

Motion components: Flexion, adduction, external rotation.

Major muscle components: Pectoralis major—clavicular portion, deltoid—anterior portion, coracobrachialis, biceps brachii—shoulder flexion component.

ELBOW

Motion components: Flexion with forearm supination.

Major muscle components: Biceps brachii—long and short heads, brachialis.

FOREARM

Motion components: Supination.
Major muscle components: Supinator.

WRIST, FINGERS, AND THUMB

See flexion–adduction–external rotation (with elbow straight) pattern.

Range limiting factors

Tension or contracture of any muscles of extension–abduction–internal rotation (with elbow extension) pattern (Fig. 16).

UPPER EXTREMITY
Flexion–Adduction–External Rotation

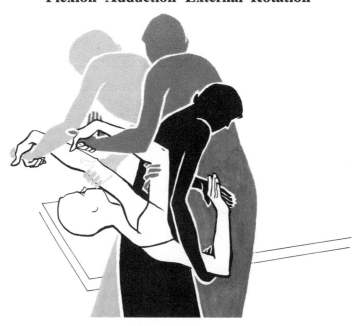

Fig. 14. With elbow extension.

Antagonistic pattern

Extension–abduction–internal rotation (with elbow flexion) (Fig. 17).

Components of motion

Fingers flex and adduct toward radial side (lateral fingers more than medial), thumb externally rotates, flexes, and adducts toward radial side, wrist supinates and flexes toward radial side, forearm supinates, elbow extends, shoulder flexes, adducts, and externally rotates with scapula rotating, abducting (inferior angle), and elevating anteriorly (acromion), and clavicle approximates sternum with rotation and elevation anteriorly.

Normal timing

Action is from distal to proximal, i.e., action occurs first at fingers, thumb, wrist, forearm, and elbow, then at shoulder, scapula, and clavicle.

Timing for emphasis

SCAPULA AND CLAVICLE

Allow beginning rotation to occur at fingers, thumb, wrist, forearm, elbow, and shoulder, but do not allow full range of finger flexion with adduction toward radial side, wrist flexion toward radial side, forearm supination, elbow extension, and shoulder flexion–adduction to occur, until scapula begins to rotate, abduct, and elevate anteriorly.

Note: If normal timing is prevented by excessive resistance to weak distal components, action cannot occur proximally. Resist stronger distal components, but guide weaker distal components through their optimal range of motion in accordance with normal timing.

SHOULDER

Allow beginning rotation to occur at fingers, thumb, wrist, forearm, elbow, and shoulder, but do not allow full range of finger flexion with adduction toward radial side, wrist flexion toward radial side, forearm supination, elbow extension, and scapular rotation to occur, until shoulder begins to flex and adduct in external rotation.

Note: Resist stronger distal components, but guide weaker distal components through their optimal range of motion in accordance with normal timing.

ELBOW

Allow beginning rotation to occur at fingers, thumb, wrist, forearm, elbow, and shoulder, but do not allow full range of finger flexion with adduction toward radial side, wrist flexion toward radial side, forearm supination, and shoulder flexion–adduction to occur, until elbow begins to extend.

Note: Resist stronger proximal and distal components, but guide weaker distal components through their optimal range of motion in accordance with normal timing.

FOREARM

Allow beginning rotation to occur at fingers, thumb, wrist, forearm, elbow, and shoulder, but do not allow full range of finger flexion with adduction toward

Patterns of facilitation

radial side, wrist flexion toward radial side, elbow extension, and shoulder flexion–adduction to occur, until forearm begins to supinate

Note: Resist stronger proximal and distal components, but guide weaker distal components through their optimal range of motion in accordance with normal timing.

WRIST

Allow beginning rotation to occur at fingers, thumb, wrist, forearm, elbow, and shoulder, but do not allow full range of finger flexion with adduction toward radial side, forearm supination, elbow extension, and shoulder flexion–adduction to occur, until wrist begins to flex toward radial side.

Note: Resist stronger proximal and distal components, but guide weaker distal components through their optimal range of motion in accordance with normal timing.

FINGERS

Allow beginning rotation to occur at fingers, thumb, wrist, forearm, elbow, and shoulder, but do not allow full range of wrist flexion toward radial side, forearm supination, elbow extension, and shoulder flexion–adduction to occur, until fingers begin to flex and adduct toward the radial side.

Note: Resist stronger proximal components. Emphasis may be placed on metacarpal-phalangeal joints, or emphasis may be placed on a specific joint of an individual digit.

THUMB

Allow beginning rotation to occur at fingers, thumb, wrist, forearm, elbow, and shoulder, but do not allow full range of finger flexion with adduction toward radial side, wrist flexion toward radial side, forearm supination, elbow extension, and shoulder flexion–adduction to occur, until thumb begins to flex and adduct. Components of fingers and wrist must be allowed to move through range after motion is initiated at thumb. In shortened range of pattern the thumb is flexed, adducted, and externally rotated toward the second metacarpal.

Note: Resist stronger finger, wrist, and proximal components.

Manual contacts

LEFT HAND

Placed in palm of patient's right hand so that patient may grasp with fingers and thumb and so that wrist may flex toward radial side (Fig. 14).

RIGHT HAND

For emphasis of distal joints: Grip with pressure of palmar surface of hand over distal anterior aspect of forearm so as to control supination and proximal components of motion.

For emphasis of elbow: Pressure of palmar surface

on anterior-medial surface of arm so as to control external rotation and proximal components of motion (Fig. 14).

For emphasis of shoulder: Same as for elbow emphasis.

For emphasis of scapula: Pressure of palmar surface of hand over the anterior aspect of shoulder proximal to acromion process.

Mass closing of hand and for emphasis of thumb motion: Grasp right thumb of patient with thumb and index finger of left hand, place right fingers and hand on palmar surface of fingers of patient's right hand. Flexion of all fingers and thumb flexion–adduction may be resisted. Physical therapist's right hand prevents range of motion occurring proximally until fingers flex and adduct.

Commands

PREPARATORY

"You are to squeeze my hand, turn it, and straighten your elbow as you pull my hand up and across your face."

ACTION

"Pull!" "Squeeze my hand!" "Turn it!" "Straighten your elbow!" "Pull it up and across your face!"

Pattern analysis

SCAPULA

Motion components: Rotation, abduction (inferior angle), elevation anteriorly (acromion).
Major muscle components: Serratus anterior.

SHOULDER

Motion components: Flexion, adduction, external rotation.
Major muscle components: Pectoralis major—clavicular portion, deltoid—anterior portion, coracobrachialis.

ELBOW

Motion components: Extension with forearm supination.
Major muscle components: Triceps—lateral portion, anconeus.

FOREARM

Motion components: Supination.
Major muscle components: Supinator.

WRIST, FINGERS, AND THUMB

See flexion–adduction–external rotation (with elbow straight) pattern.

Range limiting factors

Tension or contracture of any muscles of extension–abduction–internal rotation (with elbow flexion) pattern (Fig. 17).

UPPER EXTREMITY
Extension–Abduction–Internal Rotation

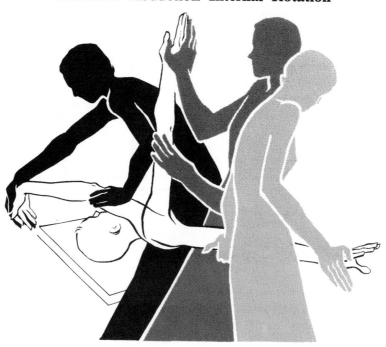

Fig. 15. With elbow straight.

Antagonistic pattern

Flexion–adduction–external rotation (with elbow straight) (Fig. 12).

Components of motion

Fingers extend and abduct toward ulnar side (medial fingers more than lateral), thumb extends, abducts, and internally rotates toward ulnar side (palmar abduction), wrist pronates and extends toward ulnar side, forearm pronates, elbow remains straight, shoulder extends, abducts, and internally rotates with scapula rotating, adducting (inferior angle) and depressing posteriorly (acromion), and clavicle rotates and depresses anteriorly away from sternum.

Normal timing

Action is from distal to proximal, i.e., action occurs first at fingers, thumb, wrist, and forearm, then at scapula, shoulder, and clavicle.

Timing for emphasis

SCAPULA AND CLAVICLE

Allow beginning rotation to occur at fingers, thumb, wrist, forearm, and shoulder, but do not allow full range of finger extension with abduction toward ulnar side, wrist pronation and extension toward ulnar side, forearm pronation, and shoulder extension–abduction to occur, until scapula begins to rotate, adduct, and depress posteriorly.

Note: If normal timing is prevented by excessive resistance to weak components, action cannot occur proximally. Resist stronger distal components, but guide weaker distal components through their optimal range of motion in accordance with normal timing.

SHOULDER

Allow beginning rotation to occur at fingers, thumb, wrist, forearm, shoulder, and scapula, but do not allow full range of finger extension with abduction toward ulnar side, wrist pronation and extension toward ulnar side, forearm pronation, and scapular rotation to occur, until shoulder begins to extend and abduct in internal rotation.

Note: Resist stronger proximal and distal components, but guide weaker distal components through their optimal range of motion in accordance with normal timing.

FOREARM

Allow beginning rotation to occur at fingers, thumb, wrist, forearm, and shoulder, but do not allow full range of finger extension with abduction toward ulnar side, wrist extension toward ulnar side, and shoulder extension–abduction to occur, until forearm begins to pronate.

Note: Resist stronger proximal and distal components, but guide weaker distal components through their optimal range of motion in accordance with normal timing.

WRIST

Allow beginning rotation to occur at fingers, thumb, wrist, forearm, and shoulder, but do not allow full range of finger extension with abduction toward ulnar side, forearm pronation, and shoulder extension–abduction to occur, until wrist begins to pronate and extend toward ulnar side.

Note: Resist stronger proximal and distal components, but guide weaker distal components through their optimal range of motion in accordance with normal timing.

FINGERS

Allow beginning rotation to occur at fingers, thumb, wrist, forearm, and shoulder, but do not allow full range of wrist pronation with extension toward ulnar side, forearm pronation, and shoulder extension-abduction to occur, until fingers begin to extend and abduct toward ulnar side.

Note: Resist stronger proximal components. Emphasis may be placed upon metacarpal-phalangeal joints, or interphalangeal joints, or emphasis may be placed upon a specific joint of an individual digit.

THUMB

Allow beginning rotation to occur at fingers, thumb, wrist, forearm, and shoulder, but do not allow full range of finger extension with abduction toward ulnar side, wrist pronation with extension toward ulnar side, forearm pronation, shoulder extension–abduction to occur, until thumb begins to extend and abduct toward ulnar side. Components of fingers and wrist must be allowed to move through range after motion is initiated at thumb. In shortened range of pattern, thumb is extended, abducted, and internally rotated away from second metacarpal.

Note: Resist stronger finger, wrist and proximal components.

Manual contacts

RIGHT HAND

Palmar surface of hand and fingers cupped over dorsal-ulnar aspect of fingers and wrist of patient's right hand (Fig. 15).

LEFT HAND

For emphasis of distal joints: Grip with pressure of palmar surface over dorsal-ulnar aspect of forearm so as to control pronation and proximal components of motion.

For emphasis of shoulder: Pressure of palmar surface on posterior-lateral surface of arm so as to control internal rotation and proximal components of motion (Fig. 15).

For emphasis of scapula: Pressure of palmar surface of hand over scapula between spine and inferior angle so as to control rotation and adduction.

For mass opening of hand and for emphasis of thumb motion: Grasp the patient's right thumb with thumb and index finger of left hand—contact of physical therapist's thumb and finger should be medially and laterally at interphalangeal joint of patient's thumb, so as to control rotation of patient's thumb as well as extension and abduction components. Physical therapist's right hand should be cupped over dorsal and ulnar aspect of patient's right hand, so as to resist wrist extension toward ulnar side and finger extension and abduction toward ulnar side. Right hand of physical therapist also controls and resists proximal components.

Commands

PREPARATORY

"You are going to open your hand, turn it, and push it down and away from your face."

ACTION

"Push!" "Open your hand!" "Turn it!" "Push it down toward me!" "Keep your elbow straight!"

Pattern analysis

SCAPULA

Motion components: Rotation, adduction (inferior angle), depression posteriorly (acromion).
Major muscle components: Levator scapulae, rhomboideii, minor and major.

SHOULDER

Motion components: Extension, abduction, internal rotation.
Major muscle components: Teres major, latissimus dorsi, deltoid—posterior portion, triceps brachii—long head (shoulder extension component).

FOREARM

Motion components: Pronation.
Major muscle components: Pronator quadratus.

WRIST

Motion components: Extension toward ulnar side.
Major muscle components: Extensor carpi ulnaris.

FINGERS

Motion components: Extension, abduction toward ulnar side.
Major muscle components: Extensor digitorum communis, extensor digiti quinti proprius, abductor digiti quinti, dorsal interossei, lumbricales.

THUMB

Motion components: Extension with abduction and rotation toward ulnar side (palmar abduction).
Major muscle components: Abductor pollicis brevis, extensor pollicis longus.

Range limiting factors

Tension or contracture of any muscles of flexion–adduction–external rotation (with elbow straight) pattern (Fig. 12).

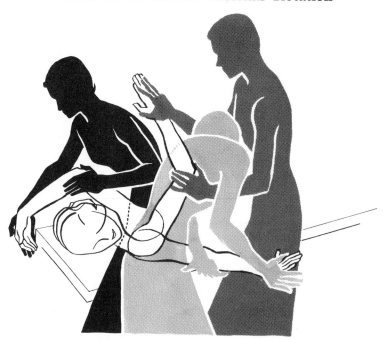

Fig. 16. With elbow extension.

Antagonistic pattern

Flexion–adduction–external rotation (with elbow flexion) (Fig. 13).

Components of motion

The fingers extend and abduct toward ulnar side (medial fingers more than lateral), thumb extends, abducts, and internally rotates toward ulnar side (palmar abduction), wrist pronates and extends toward ulnar side, forearm pronates, elbow extends, shoulder extends, abducts, and internally rotates with scapula rotating, adducting (inferior angle) and depressing posteriorly (acromion), and clavicle rotates and depresses anteriorly away from sternum.

Normal timing

Action is from distal to proximal, i.e., action occurs first at fingers, thumb, wrist, and forearm, then at elbow, scapula, shoulder, and clavicle.

Timing for emphasis

SCAPULA AND CLAVICLE

Allow beginning rotation to occur at fingers, thumb, wrist, forearm, elbow, and shoulder, but do not allow full range of finger extension with abduction toward ulnar side, wrist pronation with extension toward ulnar side, forearm pronation, elbow extension, and shoulder extension–abduction to occur, until scapula begins to rotate, adduct, and depress posteriorly.

Note: If normal timing is prevented by excessive resistance to weak components, action cannot occur proximally. Resist stronger distal components, but guide weaker distal components through their optimal range of motion in accordance with normal timing.

SHOULDER

Allow beginning rotation to occur at fingers, thumb, wrist, forearm, elbow, shoulder, and scapula, but do not allow full range of finger extension with abduction toward ulnar side, wrist pronation with extension toward ulnar side, forearm pronation, elbow extension, and scapular rotation to occur, until shoulder begins to extend and abduct in internal rotation.

Note: Resist stronger proximal and distal components, but guide weaker distal components through their optimal range of motion in accordance with normal timing.

ELBOW

Allow beginning rotation to occur at fingers, thumb, wrist, forearm, elbow, and shoulder, but do not allow full range of finger extension with abduction toward ulnar side, wrist pronation with extension toward ulnar side, forearm pronation, and shoulder extension–abduction to occur, until elbow begins to extend.

Note: Resist stronger proximal and distal components, but guide weaker distal components through their optimal range of motion in accordance with normal timing.

FOREARM

Allow beginning rotation to occur at fingers, thumb, wrist, forearm, elbow, and shoulder, but do not allow full range of finger extension with abduction toward ulnar side, wrist pronation with extension toward ulnar side, elbow extension, and shoulder extension–abduction to occur, until forearm begins to pronate.

Note: Resist stronger proximal and distal components, but guide weaker distal components through their optimal range of motion in accordance with normal timing.

WRIST

Allow beginning rotation to occur at fingers, thumb, wrist, forearm, elbow, and shoulder, but do not allow full range of finger extension with abduction toward ulnar side, forearm pronation, elbow extension, and shoulder extension–abduction to occur, until wrist begins to pronate and extend toward ulnar side.

Note: Resist stronger proximal and distal components, but guide weaker distal components through their optimal range of motion in accordance with normal timing.

FINGERS

Allow beginning rotation to occur at fingers, thumb, wrist, forearm, elbow, and shoulder, but do not allow full range of wrist pronation with extension toward ulnar side, forearm pronation, elbow extension, and shoulder extension–abduction to occur, until fingers begin to extend and abduct toward ulnar side.

Note: Resist stronger proximal components. Emphasis may be placed upon metacarpal-phalangeal joints, or interphalangeal joints, or emphasis may be placed upon a specific joint of an individual digit.

THUMB

Allow beginning rotation to occur at fingers, thumb, wrist, forearm, elbow, and shoulder, but do not allow full range of finger extension with abduction toward ulnar side, wrist pronation with extension toward ulnar side, forearm pronation, elbow extension, and shoulder extension–abduction to occur, until thumb begins to extend and abduct toward ulnar side. Components of fingers and wrist must be allowed to move through range after motion is initiated at thumb. In shortened range of pattern, thumb is extended, abducted, and internally rotated away from second metacarpal.

Note: Resist stronger finger, wrist, and proximal components.

Manual contacts

RIGHT HAND

Palmar surface of hand and fingers cupped over dorsal-ulnar aspect of fingers and wrist of patient's right hand (Fig. 16).

LEFT HAND

For emphasis of distal joints: Grip with pressure of palmar surface over dorsal-ulnar aspect of forearm, so as to control pronation and proximal components of motion.

For emphasis of shoulder and elbow: Pressure of

palmar surface on posterior-lateral surface of arm, so as to control internal rotation and proximal components of motion (Fig. 16).

For emphasis of scapula: Pressure of palmar surface of hand over scapula between spine and inferior angle, so as to control rotation and adduction.

For mass opening of hand and for emphasis of thumb motion: Grasp patient's right thumb with thumb and index finger of the left hand—contact of physical therapist's thumb and finger should be medially and laterally at interphalangeal joint of patient's thumb, so as to control rotation of patient's thumb as well as extension and abduction components. Physical therapist's right hand should be cupped over dorsal and ulnar aspect of patient's right hand, so as to resist wrist extension toward ulnar side and finger extension and abduction toward ulnar side. Right hand of physical therapist also controls and resists proximal components of extension–abduction internal rotation pattern.

Commands

PREPARATORY

"You are going to open your hand, turn it, push it down and away from your face, and extend your elbow."

ACTION

"Push!" "Open your hand!" "Turn it!" "Straighten your elbow!" "Push it down and out toward me!"

Pattern analysis

SCAPULA

Motion components: Rotation, adduction (inferior angle), depression posteriorly (acromion).
Major muscle components: Levator scapulae, rhomboidei minor and major.

SHOULDER

Motion components: Extension, abduction, internal rotation.
Major muscle components: Teres major, latissimus dorsi, deltoid—posterior portion, triceps brachii—long head (shoulder extension component).

ELBOW

Motion components: Extension.
Major muscle components: Triceps brachii—anconeus and subanconeus.

FOREARM

Motion components: Pronation.
Major muscle components: Pronator quadratus.

WRIST, FINGERS, AND THUMB

See extension–abduction–internal rotation (with elbow straight) pattern.

Range limiting factors

Tension or contracture of any muscles of the flexion–adduction–external rotation (with elbow flexion) pattern (Fig. 13).

Extension–Abduction–Internal Rotation

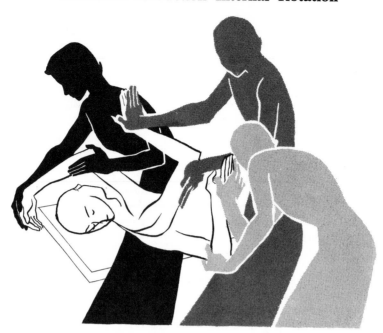

Fig. 17. With elbow flexion.

Antagonistic pattern

Flexion–adduction–external rotation (with elbow extension) (Fig. 14).

Components of motion

Fingers extend and abduct toward ulnar side (medial fingers more than lateral), thumb extends, abducts, and internally rotates toward ulnar side (palmar abduction), wrist pronates and extends toward ulnar side, forearm pronates, elbow flexes, shoulder extends, abducts, and internally rotates with scapula rotating, adducting (inferior angle), and depressing posteriorly (acromion), and clavicle rotates and depresses anteriorly away from sternum.

Normal timing

Action is from distal to proximal, i.e., action occurs first at fingers, thumb, wrist, and forearm, then at elbow, scapula, shoulder, and clavicle.

Timing for emphasis

SCAPULA AND CLAVICLE

Allow beginning rotation to occur at fingers, thumb, wrist, forearm, elbow, and shoulder, but do not allow full range of finger extension with abduction toward ulnar side, wrist pronation and extension toward ulnar side, forearm pronation, elbow flexion, and shoulder extension–abduction to occur, until scapula begins to rotate, adduct, and depress posteriorly.

Note: If normal timing is prevented by excessive resistance to weak components, action cannot occur proximally. Resist stronger distal components, but guide weaker distal components through their optimal range of motion in accordance with normal timing.

SHOULDER

Allow beginning rotation to occur at fingers, thumb, wrist, forearm, elbow, shoulder, and scapula, but do not allow full range of finger extension with abduction toward ulnar side, wrist pronation with extension toward ulnar side, forearm pronation, elbow flexion, and scapular rotation to occur, until shoulder begins to extend and abduct in internal rotation.

Note: Resist stronger proximal and distal components, but guide weaker distal components through their optimal range of motion in accordance with normal timing.

ELBOW

Allow beginning rotation to occur at fingers, thumb, wrist, forearm, elbow, and shoulder, but do not allow full range of finger extension with abduction toward ulnar side, wrist pronation with extension toward ulnar side, forearm pronation and shoulder extension–abduction to occur, until elbow begins to flex.

Note: Resist stronger proximal and distal components, but guide weaker distal components through their optimal range of motion in accordance with normal timing.

FOREARM

Allow beginning rotation to occur at fingers, thumb, wrist, forearm, elbow, and shoulder, but do not allow full range of finger extension with abduction toward ulnar side, wrist extension toward ulnar side, elbow flexion, and shoulder extension–abduction to occur, until forearm begins to pronate.

Note: Resist stronger proximal components, but guide weaker distal components through their optimal range of motion in accordance with normal timing.

WRIST

Allow beginning rotation to occur at fingers, thumb, wrist, forearm, elbow, and shoulder, but do not allow full range of finger extension with abduction toward ulnar side, forearm pronation, elbow flexion, and shoulder extension–abduction to occur, until wrist begins to pronate and extend toward ulnar side.

Note: Resist stronger proximal and distal components, but guide weaker distal components through their optimal range of motion in accordance with normal timing.

FINGERS

Allow beginning rotation to occur at fingers, thumb, wrist, forearm, elbow, and shoulder, but do not allow full range of wrist pronation with extension toward ulnar side, forearm pronation, elbow flexion, and shoulder extension–abduction to occur, until fingers begin to extend and abduct toward ulnar side.

Note: Resist stronger proximal components. Emphasis may be placed upon metacarpal-phalangeal joints, or interphalangeal joints, or emphasis may be placed upon a specific joint of an individual digit.

THUMB

Allow beginning rotation to occur at fingers, thumb, wrist, forearm, elbow, and shoulder, but do not allow full range of finger extension with abduction toward ulnar side, wrist pronation with extension toward ulnar side, forearm pronation, elbow flexion, and shoulder extension–abduction to occur, until thumb begins to extend and abduct toward ulnar side. Components of fingers and wrist must be allowed to move through range after motion is initiated at thumb. In shortened range of pattern, thumb is extended, abducted, and internally rotated away from second metacarpal.

Note: Resist stronger finger, wrist, and proximal components.

Manual contacts

RIGHT HAND

Palmar surface of hand and fingers cupped over dorsal-ulnar aspect of fingers and wrist of patient's right hand (Fig. 17).

LEFT HAND

For emphasis of distal joints: Grip with pressure of palmar surface over dorsal-ulnar aspect of forearm so as to control pronation and proximal components of motion.

For emphasis of shoulder and elbow: Pressure of palmar surface on posterior-lateral surface of arm so as to control internal rotation and proximal components of motion (Fig. 17).

For emphasis of scapula: Pressure of palmar sur-

face of hand over scapula between spine and inferior angle, so as to control rotation and adduction.

For mass opening of hand and for emphasis of thumb motion: Grasp patient's right thumb with thumb and index finger of the left hand—contact of physical therapist's thumb and finger should be medially and laterally at interphalangeal joint of patient's thumb, so as to control rotation of patient's thumb as well as extension and abduction components. Physical therapist's right hand should be cupped over dorsal and ulnar aspect of patient's right hand, so as to resist wrist extension toward ulnar side and finger extension and abduction toward ulnar side. Right hand of physical therapist also controls and resists proximal components of extension–abduction–internal rotation pattern.

Commands

PREPARATORY

"You are going to open your hand, turn it, push it down and away from your face, and bend your elbow."

ACTION

"Push!" "Open your hand!" "Turn it!" "Bend your elbow!" "Push it toward me!"

Pattern analysis*

SCAPULA

Motion components: Rotation, adduction (inferior angle), depression posteriorly (acromion).

Major muscle components: Levator scapulae, rhomboidei minor and major.

SHOULDER

Motion components: Extension, abduction, internal rotation.

Major muscle components: Teres major, latissimus dorsi, deltoid—posterior portion.

ELBOW

Motion components: Flexion.

Major muscle components: Brachialis, biceps brachii—lateral portion.

FOREARM

Motion components: Pronation.

Major muscle components: Pronator quadratus.

WRIST, FINGERS, AND THUMB

See extension–abduction–internal rotation (with elbow straight) pattern.

RANGE LIMITING FACTORS

Tension or contracture of any muscles of the flexion–adduction–external rotation (with elbow extension) pattern (Fig. 14).

---

* If extension–abduction–internal rotation is continued posteriorly, the motion components combine with those of the extension–adduction–internal rotation pattern. The fingers and wrist flex toward the ulnar side, the elbow (may remain straight or flex) and the shoulder extends, adducts and internally rotates with the scapula rotating, adducting (inferior angle) and depressing anteriorly (acromion process). This movement is important to the complete reeducation of the latissimus dorsi with consideration given to the adduction component of this muscle.

UPPER EXTREMITY
Flexion–Abduction–External Rotation

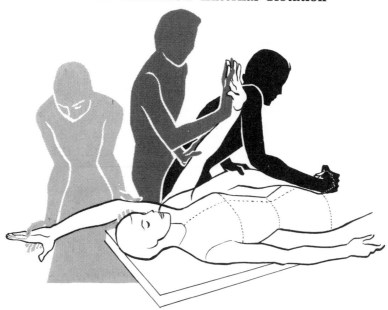

Fig. 18. With elbow straight.

Antagonistic pattern

Extension–adduction–internal rotation (with elbow straight) (Fig. 21).

Components of motion

Fingers extend and abduct toward radial side (lateral fingers more than medial) thumb extends, adducts, and externally rotates toward radial side, wrist supinates and extends toward radial side, forearm supinates, elbow remains straight, shoulder flexes, abducts, and externally rotates with scapula rotating, adducting (medial angle), and elevating posteriorly (acromion), and clavicle rotates and elevates anteriorly away from sternum.

Normal timing

Action is from distal to proximal, i.e., action occurs first at fingers, thumb, wrist, and forearm, then at scapula, shoulder, and clavicle.

Timing for emphasis

SCAPULA AND CLAVICLE

Allow beginning rotation to occur at fingers, thumb, wrist, forearm, and shoulder, but do not allow full range of finger extension with abduction toward radial side, wrist supination with extension toward radial side, forearm supination, and shoulder flexion–abduction to occur, until scapula begins to rotate, adduct, and elevate posteriorly.

Note: If normal timing is prevented by excessive resistance to weak components, action cannot occur proximally. Resist stronger distal components, but guide weaker distal components through their optimal range of motion in accordance with normal timing.

SHOULDER

Allow beginning rotation to occur at fingers, thumb, wrist, forearm, shoulder, and scapula, but do not allow full range of finger extension with abduction toward radial side, wrist supination with extension toward radial side, forearm supination, and scapular rotation to occur, until shoulder begins to flex and abduct in external rotation.

Note: Resist stronger proximal and distal components, but guide weaker distal components through their optimal range of motion in accordance with normal timing.

FOREARM

Allow beginning rotation to occur at fingers, thumb, wrist, forearm, and shoulder, but do not allow full range of finger extension with abduction toward radial side, wrist supination with extension toward radial side, and shoulder flexion–abduction to occur, until forearm begins to supinate.

Note: Resist stronger proximal and distal components, but guide weaker distal components through their optimal range of motion in accordance with normal timing.

WRIST

Allow beginning rotation to occur at fingers, thumb, wrist, forearm and shoulder, but do not allow full range of finger extension with abduction toward radial side, forearm supination, and shoulder flexion–abduction to occur, until wrist begins to supinate and extend toward radial side.

Note: Resist stronger proximal and distal components, but guide weaker distal components through their optimal range of motion in accordance with normal timing.

FINGERS

Allow beginning rotation to occur at fingers, thumb, wrist, forearm, and shoulder, but do not allow full range of wrist supination with extension toward radial side, forearm supination, and shoulder flexion–abduction to occur, until fingers begin to extend and abduct toward radial side.

Note: Resist stronger proximal components. Emphasis may be placed upon metacarpal-phalangeal joints, or upon interphalangeal joints, or emphasis may be placed upon a specific joint of an individual digit.

THUMB

Allow beginning rotation to occur at fingers, thumb, wrist, forearm, and shoulder, but do not allow full range of finger extension with abduction, wrist supination with extension toward radial side forearm supination, shoulder flexion–abduction to occur, until thumb begins to extend and adduct toward radial side. Components of fingers and wrist must be allowed to move through range after motion is initiated at thumb. In shortened range of pattern thumb is extended, adducted, and externally rotated toward second metacarpal.

Note: Resist stronger finger, wrist, and proximal components.

Manual contacts
RIGHT HAND

Palmar surface of hand and fingers cupped over dorsal-radial aspect of fingers and wrist of patient's left hand (Fig. 18).

LEFT HAND

For emphasis of distal joints: Grip with pressure of palmar surface over dorsal-radial aspect of forearm, so as to control supination and proximal components of motion.

For emphasis of shoulder: Pressure of palmar surface on anterior-lateral surface of patient's arm, so as to control rotation and proximal components of motion (Fig. 18).

For emphasis of scapula: Pressure of palmar surface of hand over scapula at medial angle, so as to control rotation and adduction.

For mass opening of hand and for emphasis of thumb motion: Grasp patients' left thumb with thumb and index finger of right hand—contact of physical therapist's thumb and finger should be medially and laterally at interphalangeal joint of patient's thumb,

so as to control rotation of patient's thumb as well as extension and adduction components. Physical therapist's left hand should be cupped over dorsal and radial aspect of patient's left hand, so as to resist wrist extension toward radial side and finger extension with abduction toward radial side. Right hand of physical therapist also controls and resists proximal components of flexion–abduction–external rotation pattern.

Commands
PREPARATORY

"You are going to open your hand, turn it, and lift it up and out toward me, keeping your elbow straight."

ACTION

"Lift!" "Open your hand!" "Turn it!" "Keep your elbow straight!" "Lift it up toward me!"

Pattern analysis
SCAPULA

Motion components: Rotation, adduction (medial angle), elevation posteriorly (acromion).
Major muscle components: Trapezius—upper, middle, and lower portions.

SHOULDER

Motion components: Flexion, abduction, external rotation.
Major muscle components: Teres minor, supraspinatus, infraspinatus, deltoid—middle portion.

FOREARM

Motion components: Supination.
Major muscle components: Brachioradialis.

WRIST

Motion components: Extension toward radial side.
Major muscle components: Extensor carpi radialis longus, extensor carpi radialis brevis.

FINGERS

Motion components: Extension, abduction toward radial side.
Major muscle components: Extensor digitorum communis, extensor indicis proprius, dorsal interossei, lumbricales.

THUMB

Motion components: Extension with adduction and rotation toward radial side.
Major muscle components: Extensor pollicis longus, abductor pollicis longus, extensor pollicis brevis, first dorsal interosseus.

Range limiting factors

Tension or contracture of any muscles of extension–adduction–internal rotation (with elbow straight) pattern (Fig. 21).

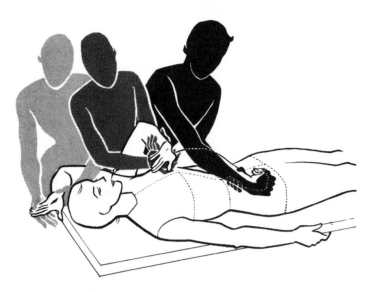

Fig. 19. With elbow flexion.

Antagonistic pattern

Extension–adduction–internal rotation (with elbow extension) (Fig. 22).

Components of motion

Fingers extend and abduct toward radial side (lateral fingers more than medial) thumb extends, adducts, and externally rotates toward radial side, wrist supinates and extends toward radial side, forearm supinates, elbow flexes, shoulder flexes, abducts, and externally rotates with scapula rotating, adducting (medial angle), and elevating posteriorly (acromion), and clavicle rotates and elevates anteriorly away from sternum.

Normal timing

Action is from distal to proximal, i.e., action occurs first at fingers, thumb, wrist, and forearm, then at elbow, scapula, shoulder, and clavicle.

Timing for emphasis

SCAPULA AND CLAVICLE

Allow beginning rotation to occur at fingers, thumb, wrist, forearm, elbow, and shoulder, but do not allow full range of finger extension with abduction toward radial side, wrist supination with extension toward the radial side, forearm supination, elbow flexion, and shoulder flexion–abduction to occur, until scapula begins to rotate, adduct, and elevate posteriorly.

Note: If normal timing of pattern is prevented by excessive resistance to weak components, action cannot occur proximally. Resist stronger distal components, but guide weaker distal components through their optimal range of motion in accordance with normal timing.

SHOULDER

Allow beginning rotation to occur at fingers, thumb, wrist, forearm, elbow, shoulder, and scapula, but do not allow full range of finger extension with abduction toward radial side, wrist supination with extension toward radial side, forearm supination, elbow flexion, and scapular rotation to occur, until shoulder begins to flex and abduct in external rotation.

Note: Resist stronger proximal and distal components, but guide weaker distal components through their optimal range of motion in accordance with normal timing.

ELBOW

Allow beginning rotation to occur at fingers, thumb, wrist, forearm, elbow, and shoulder, but do not allow full range of finger extension with abduction toward radial side, wrist supination with extension toward radial side, forearm supination, and shoulder flexion-abduction to occur, until elbow begins to flex.

Note: Resist stronger proximal and distal components, but guide weaker distal components through their optimal range of motion in accordance with normal timing.

FOREARM

Allow beginning rotation to occur at fingers, thumb, wrist, forearm, elbow, and shoulder, but do not allow full range of finger extension with abduction toward radial side, wrist supination with extension toward radial side, elbow flexion, and shoulder flexion–abduction to occur, until forearm beings to supinate.

Note: Resist stronger proximal and distal components, but guide weaker distal components through their optimal range of motion in accordance with normal timing.

WRIST

Allow beginning rotation to occur at fingers, thumb, wrist, forearm, elbow, and shoulder, but do not allow full range of finger extension with abduction toward radial side, forearm supination, elbow flexion, and shoulder flexion–abduction to occur, until wrist begins to supinate and extend toward radial side.

Note: Resist stronger proximal and distal components, but guide weaker components through their optimal range of motion in accordance with normal timing.

FINGERS

Allow beginning rotation to occur at fingers, thumb, wrist, forearm, elbow, and shoulder, but do not allow full range of wrist supination with extension toward radial side, forearm supination, elbow flexion, and shoulder flexion–abduction to occur, until fingers begin to extend and abduct toward radial side.

Note: Resist stronger proximal components. Emphasis may be placed upon metacarpal-phalangeal joints, or interphalangeal joints, or emphasis may be placed upon a specific joint of an individual digit.

THUMB

Allow beginning rotation to occur at fingers, thumb, wrist, forearm, elbow, and shoulder, but do not allow full range of finger extension with abduction toward radial side, wrist supination with extension toward radial side, forearm supination, elbow flexion, and shoulder flexion–abduction to occur, until thumb begins to extend and adduct toward radial side. Components of fingers and wrist must be allowed to move through range after motion is initiated at thumb. In shortened range of pattern, thumb is extended, adducted, and externally rotated toward second metacarpal.

Note: Resist stronger finger, wrist, and proximal components.

Manual contacts
RIGHT HAND

Palmar surface of hand and fingers cupped over dorsal-radial aspect of fingers and wrist of patient's left hand (Fig. 19).

LEFT HAND

For emphasis of distal joints: Grip with pressure of palmar surface over dorsal-radial aspect of forearm, so as to control supination and proximal components of motion (Fig. 19).

For emphasis of shoulder and elbow: Pressure of palmar surface on anterior-lateral surface of patient's arm, so as to control rotation and proximal components of motion.

For emphasis of scapula: Pressure of palmar surface of hand over scapula at medial angle, so as to control rotation and adduction.

For mass opening of hand and for emphasis of thumb motion: Grasp patient's left thumb with thumb and index finger of right hand—contact of physical therapist's thumb and finger should be medially and laterally at interphalangeal joint of patient's thumb, so as to control rotation of patient's thumb as well as extension and adduction components. Physical therapist's left hand should be cupped over dorsal and radial aspect of patient's left hand, so as to resist extension toward radial side and finger extension with abduction toward radial side. Right hand of physical therapist also controls and resists proximal components of flexion–abduction–external rotation pattern.

Commands
PREPARATORY

"You are going to open your hand, turn it, and lift it up and out toward me, bending your elbow."

ACTION

"Pull!" "Open your hand!" "Turn it!" "Bend your elbow!" "Pull it up here toward me!"

Pattern analysis
SCAPULA

Motion components: Rotation, adduction (medial angle), elevation posteriorly (acromion).
Major muscle components: Trapezius—upper, middle, and lower portions.

SHOULDER

Motion components: Flexion, abduction, external rotation.
Major muscle components: Teres minor, supraspinatus, infraspinatus, deltoid—middle portion, biceps brachii (long head—shoulder flexion component).

ELBOW

Motion components: Flexion.
Major muscle components: Biceps brachii—long head (lateral portion), brachioradialis.

FOREARM

Motion components: Supination.
Major muscle components: Brachioradialis.

WRIST, FINGERS, AND THUMB

See flexion–abduction–external rotation (with elbow straight) pattern.

Range limiting factors

Tension or contracture of any muscles of extension–adduction–internal rotation (with elbow extension) pattern (Fig. 22).

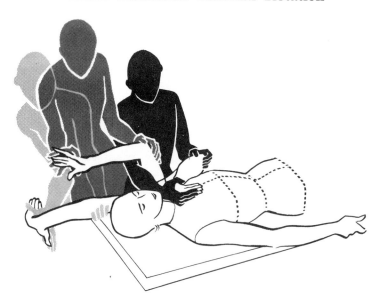

Fig. 20. With elbow extension.

Antagonistic pattern

Extension–adduction–internal rotation (with elbow flexion) (Fig. 23).

Components of motion

Fingers extend with abduction toward radial side (lateral fingers more than medial), thumb extends, adducts, and externally rotates toward radial side, wrist supinates and extends toward radial side, forearm supinates, elbow extends, shoulder flexes, abducts, and externally rotates with scapula rotating, adducting (medial angle), and elevating posteriorly (acromion), and clavicle rotates and elevates anteriorly away from sternum.

Normal timing

Action occurs from distal to proximal, i.e., action occurs first at fingers, thumb, wrist, and forearm, then at elbow, scapula, shoulder, and clavicle.

Timing for emphasis

SCAPULA AND CLAVICLE

Allow beginning rotation to occur at fingers, thumb, wrist, forearm, elbow, and shoulder, but do not allow full range of finger extension with abduction toward radial side, wrist supination with extension toward radial side, forearm supination, elbow extension, and shoulder flexion–abduction to occur, until scapula begins to rotate, adduct, and elevate posteriorly.

Note: If normal timing is prevented by excessive resistance to weak components, action cannot occur proximally. Resist stronger distal components, but guide weaker distal components through their optimal range of motion in accordance with normal timing.

SHOULDER

Allow beginning rotation to occur at fingers, thumb, wrist, forearm, elbow, shoulder, and scapula, but do not allow full range of finger extension with abduction toward radial side, wrist supination with extension toward radial side, forearm supination, elbow extension, and scapular rotation to occur, until shoulder begins to flex and abduct in external rotation.

Note: Resist stronger proximal and distal components, but guide weaker distal components through their optimal range of motion in accordance with normal timing.

ELBOW

Allow beginning rotation to occur at fingers, thumb, wrist, forearm, elbow, and shoulder, but do not allow full range of finger extension with abduction toward radial side, wrist supination with extension toward radial side, forearm supination, and shoulder flexion–abduction to occur, until elbow begins to flex.

Note: Resist stronger proximal and distal components but guide weaker distal components through their optimal range of motion in accordance with normal timing.

FOREARM

Allow beginning rotation to occur at fingers, thumb, wrist, forearm, elbow, and shoulder, but do not allow full range of finger extension with abduction toward radial side, wrist supination with extension toward radial side, elbow extension, and shoulder flexion–abduction to occur, until forearm begins to supinate.

Note: Resist stronger proximal and distal components, but guide weaker distal components through their optimal range of motion in accordance with normal timing.

WRIST

Allow beginning rotation to occur at fingers, thumb, wrist, forearm, elbow, and shoulder, but do not allow full range of finger extension with abduction toward radial side, forearm supination, elbow extension, and shoulder flexion–abduction to occur, until wrist begins to supinate and extend toward radial side.

Note: Resist stronger proximal and distal components, but guide weaker distal components through their optimal range of motion in accordance with normal timing.

FINGERS

Allow beginning rotation to occur at fingers, thumb, wrist, forearm, elbow, and shoulder, but do not allow full range of wrist supination with extension toward radial side, forearm supination, elbow extension, and shoulder flexion–abduction to occur, until the fingers begin to extend and abduct toward radial side.

Note: Resist stronger proximal components. Emphasis may be placed upon metacarpal-phalangeal joints, or interphalangeal joints, or emphasis may be placed upon a specific joint of an individual digit.

THUMB

Allow beginning rotation to occur at fingers, thumb, wrist, forearm, elbow, and shoulder, but do not allow full range of finger extension with abduction toward radial side, wrist supination with extension toward radial side, forearm supination, elbow extension, and shoulder flexion–abduction to occur, until thumb begins to extend and adduct toward radial side. Components of fingers and wrist must be allowed to move through range after motion is initiated at thumb. In shortened range of pattern thumb is extended, adducted, and externally rotated toward second metacarpal.

Note: Resist stronger finger, wrist, and proximal components.

Manual contacts

RIGHT HAND

Palmar surface of hand and fingers cupped over dorsal-radial aspect of fingers and wrist of patient's left hand (Fig. 20).

LEFT HAND

For emphasis of distal joints: Grip with pressure of palmar surface over dorsal-radial aspect of forearm, so as to control supination and proximal components of motion.

For emphasis of shoulder and elbow: Pressure of palmar surface on anterior-lateral surface of patient's arm, so as to control rotation and proximal components of motion (Fig. 20).

For emphasis of scapula: Pressure of palmar surface of hand over scapula at medial angle, so as to control rotation and adduction.

For mass opening of hand and for emphasis of thumb motion: Grasp patient's left thumb with thumb and index finger of right hand—contact of physical therapist's thumb and finger should be medially and laterally at interphalangeal joint of patient's thumb, so as to control rotation of patient's thumb as well as extension and adduction components. Physical therapist's left hand should be cupped over dorsal and radial aspect of patient's left hand, so as to resist wrist extension toward radial side and finger extension with abduction toward radial side. Right hand of physical therapist also controls and resists proximal components of flexion–abduction–external rotation pattern.

Commands

PREPARATORY

"You are going to open your hand, turn it, and push it up and out toward me, straightening your elbow."

ACTION

"Push!" "Open your hand!" "Turn it!" "Push it up and out toward me!" "Straighten your elbow!"

Pattern analysis

SCAPULA

Motion components: Rotation, adduction (medial angle), elevation posteriorly (acromion).

Major muscle components: Trapezius—upper, middle and lower portions.

SHOULDER

Motion components: Flexion, abduction, external rotation.

Major muscle components: Teres minor, supraspinatus, infraspinatus, deltoid—middle portion.

ELBOW

Motion components: Extension.

Major muscle components: Triceps brachii—lateral portion, anconeus.

FOREARM

Motion components: Supination.

Major muscle components: Brachioradialis.

WRIST, FINGERS, AND THUMB

See flexion–abduction–external rotation (with elbow straight) pattern.

Range limiting factors

Tension or contracture of any muscles of extension–adduction–internal rotation (with elbow flexion) pattern (Fig. 23).

UPPER EXTREMITY
Extension–Adduction–Internal Rotation

Fig. 21. With elbow straight.

Antagonistic pattern

Flexion–abduction–external rotation (with elbow straight) (Fig. 18).

Components of motion

Fingers flex and adduct (medial fingers more than lateral) toward ulnar side, thumb flexes, abducts, and internally rotates toward ulnar side (opposition), wrist pronates and flexes toward ulnar side, forearm pronates, elbow remains straight, shoulder extends, adducts, and internally rotates with scapula rotating, abducting (medial angle) and depressing anteriorly (acromion), and clavicle rotates and depresses anteriorly in approximation with sternum.

Normal timing

Action is from distal to proximal, i.e., action occurs first at fingers, thumb, wrist, and forearm, then at shoulder, scapula, and clavicle.

Timing for emphasis
SCAPULA AND CLAVICLE

Allow beginning rotation to occur at fingers, thumb, wrist, forearm, and shoulder, but do not allow full range of finger flexion with adduction toward ulnar side, wrist pronation and flexion toward ulnar side, forearm pronation, and shoulder extension–adduction to occur, until scapula begins to rotate, abduct, and depress anteriorly.

Note: If normal timing is prevented by excessive resistance to weak components, action cannot occur proximally. Resist stronger distal components, but guide weaker components through their optimal range of motion in accordance with normal timing.

SHOULDER

Allow beginning rotation to occur at fingers, thumb, wrist, forearm, and shoulder, but do not allow full range of finger flexion with adduction toward ulnar side, wrist pronation with flexion toward ulnar side, and forearm pronation to occur, until shoulder begins to extend and adduct in internal rotation.

Note: Resist stronger distal components, but guide weaker components through their optimal range of motion in accordance with normal timing.

FOREARM

Allow beginning rotation to occur at fingers, thumb, wrist, forearm, and shoulder, but do not allow full range of finger flexion with adduction toward ulnar side, wrist pronation with flexion toward ulnar side, and shoulder extension–adduction to occur, until forearm begins to pronate.

Note: Resist stronger proximal and distal components, but guide weaker distal components through their optimal range of motion in accordance with normal timing.

WRIST

Allow beginning rotation to occur at fingers, thumb, wrist, forearm, and shoulder, but do not allow full range of finger flexion with adduction toward ulnar side, forearm pronation, and shoulder extension–adduc-

Patterns of facilitation

tion to occur, until wrist begins to pronate and flex toward ulnar side.

Note: Resist stronger proximal and distal components, but guide weaker distal components through their optimal range of motion in accordance with normal timing.

FINGERS

Allow beginning rotation to occur at fingers, thumb, wrist, forearm, and shoulder, but do not allow full range of wrist pronation with flexion toward ulnar side, forearm pronation, and shoulder extension–adduction to occur, until fingers begin to flex and adduct toward ulnar side.

Note: Resist stronger proximal components. Emphasis may be placed upon metacarpal-phalangeal joints, or interphalangeal joints, or emphasis may be placed upon a specific joint of an individaul digit.

THUMB

Allow beginning rotation to occur at fingers, thumb, wrist, forearm, and shoulder, but do not allow full range of finger flexion with adduction toward ulnar side, wrist pronation with flexion toward ulnar side, forearm pronation, and shoulder extension–adduction to occur, until thumb begins to flex and abduct toward ulnar side. Components of fingers and wrist must be allowed to move through range after motion is initiated at thumb. In shortened range of pattern thumb is flexed, abducted and internally rotated away from second metacarpal and toward fifth metacarpal.

Note: Resist stronger components of fingers and wrist and all proximal components.

Manual contacts

LEFT HAND

Placed in palm of patient's left hand so that patient may grasp with fingers and thumb, and so that wrist may flex toward ulnar side (Fig. 21).

RIGHT HAND

For emphasis of distal joints: Grip with pressure of palmar surface over anterior-ulnar aspect of patient's forearm, so as to control pronation and proximal components of motion (Fig. 21).

For emphasis of shoulder: Pressure of palmar surface on posterior-medial surface of patient's arm, so as to control internal rotation and proximal components of motion.

For emphasis of scapula: Pressure of palmar surface of hand over anterior-medial axillary space and acromion process.

For mass closing of hand and for emphasis of thumb motion: Grasp patient's left thumb with thumb and index finger of right hand—contact of physical therapist's thumb and index finger should be medially and laterally at interphalangeal joint of patient's thumb, so as to control rotation of patient's thumb as well as the

flexion and abduction components. Physical therapist's left hand should be placed with palmar surface of hand and fingers on palmar surface of hand and fingers of patient's left hand. Physical therapist's left hand prevents range of motion occurring proximally until fingers flex and adduct toward ulnar side. Flexion of all fingers and thumb flexion–adduction may be resisted.

Commands

PREPARATORY

"You are going to squeeze my hand, turn it, and pull it down toward your right hip, keeping your elbow straight."

ACTION

"Pull!" "Squeeze my hand!" "Turn it!" "Keep your elbow straight!" "Pull it down toward your right hip!"

Pattern analysis

SCAPULA

Motion components: Rotation, abduction (medial angle), depression anteriorly (acromion).

Major muscle components: Pectoralis minor, subclavius—acting upon clavicle.

SHOULDER

Motion components: Extension, adduction, internal rotation.

Major muscle components: Subscapularis, pectoralis major—sternal portion.

FOREARM

Motion components: Pronation.

Major muscle components: Pronator teres.

WRIST

Motion components: Pronation, flexion toward ulnar side.

Major muscle components: Flexor carpi ulnaris, palmaris longus.

FINGERS

Motion components: Flexion, adduction toward ulnar side.

Major muscle components: Flexor digitorum superficialis, flexor digitorum profundus, palmar interossei, lumbricales.

THUMB

Motion components: Flexion, abduction, rotation away from second metacarpal.

Major muscle components: Flexor pollicis longus, flexor pollicis brevis, opponens pollicis, palmaris brevis.

Range limiting factors

Tension or contracture of any muscles of the flexion–abduction–external rotation (with elbow straight) pattern (Fig. 18).

UPPER EXTREMITY
Extension–Adduction–Internal Rotation

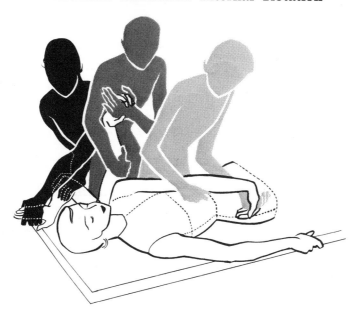

Fig. 22. With elbow extension.

Antagonistic pattern

Flexion–abduction–external rotation (with elbow flexion) (Fig. 19).

Components of motion

Fingers flex and adduct (medial fingers more than lateral) toward ulnar side, thumb flexes, abducts, and internally rotates toward ulnar side (opposition), wrist pronates and flexes toward ulnar side, forearm pronates, elbow extends, shoulder extends, abducts, and internally rotates with scapula rotating, abducting (medial angle), and depressing anteriorly (acromion), and clavicle rotates and depresses anteriorly in approximation with sternum.

Normal timing

Action is from distal to proximal, i.e., action occurs first at fingers, thumb, wrist, and forearm, then at elbow, shoulder, scapula, and clavicle.

Timing for emphasis
SCAPULA AND CLAVICLE

Allow beginning rotation to occur at fingers, thumb, wrist, forearm, elbow, and shoulder, but do not allow full range of finger flexion with adduction toward ulnar side, wrist pronation with flexion toward ulnar side, forearm pronation, elbow extension, and shoulder extension–adduction to occur, until scapula begins to rotate, abduct, and depress anteriorly.

Note: If the normal timing is prevented by excessive resistance to weak components, action cannot occur proximally. Resist stronger distal components, but guide weaker distal components through their optimal range of motion in accordance with normal timing.

SHOULDER

Allow beginning rotation to occur at fingers, thumb, wrist, forearm, elbow, and shoulder, but do not allow full range of finger flexion with adduction toward ulnar side, wrist pronation with flexion toward ulnar side, forearm pronation, and elbow extension to occur, until shoulder begins to extend and adduct in internal rotation.

Note: Resist stronger distal components, but guide weaker components through their optimal range.

ELBOW

Allow beginning rotation to occur at fingers, thumb, wrist, forearm, elbow, and shoulder, but do not allow full range of finger flexion with adduction toward ulnar side, wrist pronation with flexion toward ulnar side, forearm pronation, and shoulder extension–adduction to occur, until elbow begins to extend.

Note: Resist stronger proximal and distal components, but guide weaker distal components through their optimal range of motion in accordance with normal timing.

FOREARM

Allow beginning rotation to occur at fingers, thumb, wrist, forearm, elbow, and shoulder, but do not allow full range of finger flexion with adduction toward ulnar side, wrist pronation with flexion toward ulnar side, elbow extension, and shoulder extension–adduction to occur, until forearm begins to pronate.

Patterns of facilitation

Note: Resist stronger proximal and distal components, but guide weaker distal components through their optimal range of motion in accordance with normal timing.

WRIST

Allow beginning rotation to occur at fingers, thumb, wrist, forearm, elbow, and shoulder, but do not allow full range of finger flexion with adduction toward ulnar side, forearm pronation, elbow extension, and shoulder extension–adduction to occur, until wrist begins to pronate and flex toward ulnar side.

Note: Resist stronger proximal and distal components, but guide weaker distal components through their optimal range of motion in accordance with normal timing.

FINGERS

Allow beginning rotation to occur at fingers, thumb, wrist, forearm, elbow, and shoulder, but do not allow full range of wrist pronation with flexion toward ulnar side, forearm pronation, elbow extension, and shoulder extension–adduction to occur, until fingers begin to flex and adduct toward ulnar side.

Note: Resist all stronger proximal components. Emphasis may be placed upon metacarpal-phalangeal joints, or interphalangeal joints, or emphasis may be placed upon a specific joint of an individual digit.

THUMB

Allow beginning rotation to occur at fingers, thumb, wrist, forearm, elbow, and shoulder, but do not allow full range of finger flexion with adduction toward ulnar side, wrist pronation with flexion toward ulnar side, forearm pronation, elbow extension, and shoulder extension–adduction to occur, until thumb begins to flex and abduct toward ulnar side. Components of finger and wrist must be allowed to move through range after motion is initiated at thumb. In shortened range of pattern thumb is flexed, abducted, and internally rotated away from second metacarpal and toward fifth metacarpal.

Note: Resist all stronger finger, wrist, and proximal components.

Manual contacts

LEFT HAND

Placed in palm of patient's left hand, so that patient may grasp with fingers and thumb, and so that wrist may flex toward ulnar side (Fig. 22).

RIGHT HAND

For emphasis of distal joints: Grip with pressure of palmar surface over anterior-ulnar aspect of patient's forearm, so as to control pronation and proximal components of motion (Fig. 22).

For emphasis of shoulder and elbow: Pressure of palmar surface on posterior-medial surface of patient's arm, so as to control internal rotation and proximal components of motion.

For emphasis of scapula: Pressure of palmar surface of hand over anterior-medial axillary space and acromion process.

For mass closing of hand and for emphasis of thumb motion: Grasp patient's left thumb with thumb and index finger of right hand—contact of physical therapist's thumb and index finger should be medially and laterally at interphalangeal joint of patient's thumb, so as to control rotation of patient's thumb as well as flexion and abduction components. Physical therapist's left hand should be placed with palmar surface of hand and fingers on palmar surface of hand and fingers of patient's left hand. Physical therapist's left hand prevents range of motion occurring proximally until fingers flex and adduct toward ulnar side. Flexion of all fingers and thumb flexion–abduction may be resisted.

Commands

PREPARATORY

"You are to squeeze my hand, turn it, and push it down and toward your right hip, straightening your elbow."

ACTION

"Push!" "Squeeze my hand!" "Turn it!" "Straighten your elbow!" "Push it toward your right hip!"

Pattern analysis

SCAPULA

Motion components: Rotation, abduction (medial angle), depression anteriorly (acromion).

Major muscle components: Pectoralis minor, subclavius—acting upon clavicle.

SHOULDER

Motion components: Extension, adduction, internal rotation.

Major muscle components: Subscapularis, pectoralis major—sternal portion, triceps brachii—long head (shoulder extension component).

ELBOW

Motion components: Extension.

Major muscle components: Triceps brachii, anconeus, and subanconeus.

FOREARM

Motion components: Pronation.

Major muscle components: Pronator teres.

WRIST, FINGERS, AND THUMB

See extension–adduction–internal rotation (with elbow straight) pattern.

Range limiting factors

Tension or contracture of any muscles of the flexion–abduction–external rotation (with elbow flexion) pattern (Fig. 19).

UPPER EXTREMITY
Extension–Adduction–Internal Rotation

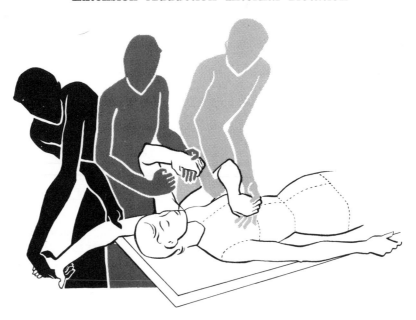

Fig. 23. With elbow flexion.

Antagonistic pattern

Flexion–abduction–external rotation (with elbow extension) (Fig. 20).

Components of motion

Fingers flex and adduct (medial fingers more than lateral) toward ulnar side, thumb flexes, abducts, and internally rotates toward ulnar side (opposition), wrist pronates and flexes toward ulnar side, forearm pronates, elbow flexes, shoulder extends, adducts, and internally rotates with scapula rotating, abducting (medial angle), and depressing anteriorly (acromion), and clavicle rotates and depresses anteriorly in approximation with sternum.

Normal timing

Action is from distal to proximal, i.e., action occurs first at fingers, thumb, wrist, and forearm, then at elbow, shoulder, scapula, and clavicle.

Timing for emphasis

SCAPULA AND CLAVICLE

Allow beginning rotation to occur at fingers, thumb, wrist, forearm, elbow, and shoulder, but do not allow full range of finger flexion with adduction toward ulnar side, wrist pronation with flexion toward ulnar side, forearm pronation, elbow flexion, and shoulder extension–adduction to occur, until scapula begins to rotate, abduct, and depress anteriorly.

Note: If normal timing is prevented by excessive resistance to weak components, action cannot occur proximally. Resist stronger distal components, but guide weaker distal components through their optimal range of motion in accordance with normal timing.

SHOULDER

Allow beginning rotation to occur at fingers, thumb, wrist, forearm, elbow, and shoulder, but do not allow full range of finger flexion with adduction toward ulnar side, wrist pronation with flexion toward ulnar side, forearm pronation, and elbow flexion to occur, until shoulder begins to extend and adduct in internal rotation.

Note: Resist stronger distal components, but guide weaker distal components through their optimal range of motion in accordance with normal timing.

ELBOW

Allow beginnning rotation to occur at fingers, thumb, wrist, forearm, elbow, and shoulder, but do not allow full range of finger flexion with adduction toward ulnar side, wrist pronation with flexion toward ulnar side, forearm pronation, and shoulder extension–adduction to occur, until elbow begins to flex.

Note: Resist stronger proximal and distal components, but guide weaker distal components through their optimal range of motion in accordance with normal timing.

FOREARM

Allow beginning rotation to occur at fingers, thumb, wrist, forearm, elbow, and shoulder, but do not allow full range of finger flexion with adduction toward ulnar side, wrist pronation with flexion toward ulnar side,

Patterns of facilitation

elbow flexion, and shoulder extension–adduction to occur, until forearm begins to pronate.

Note: Resist stronger proximal and distal components, but guide weaker distal components through their optimal range of motion in accordance with normal timing.

WRIST

Allow beginning rotation to occur at fingers, thumb, wrist, forearm, elbow, and shoulder, but do not allow full range of finger flexion with adduction toward ulnar side, forearm pronation, elbow flexion, and shoulder extension–adduction to occur, until wrist begins to pronate and flex toward ulnar side.

Note: Resist stronger proximal and distal components, but guide weaker distal components through their optimal range of motion in accordance with normal timing.

FINGERS

Allow beginning rotation to occur at fingers, thumb, wrist, forearm, elbow, and shoulder, but do not allow full range of wrist pronation with flexion toward ulnar side, forearm pronation, elbow flexion, and shoulder extension–adduction to occur, until fingers begin to flex and adduct toward ulnar side.

Note: Resist stronger proximal and distal components, but guide weaker distal components through their optimal range of motion in accordance with normal timing. Emphasis may be placed upon metacarpal-phalangeal joints, or interphalangeal joints, or emphasis may be placed upon a specific joint of an individual digit.

THUMB

Allow beginning rotation to occur at fingers, thumb, wrist, forearm, elbow, and shoulder, but do not allow full range of finger flexion with adduction toward ulnar side, wrist pronation with flexion toward ulnar side, forearm pronation, elbow flexion, and shoulder extension–adduction to occur, until thumb begins to flex and abduct toward ulnar side. Components of fingers and wrist must be allowed to move through range after motion is initiated at thumb. In shortened range of pattern, thumb is flexed, abducted, and internally rotated away from second metacarpal and toward fifth metacarpal.

Note: Resist stronger components of fingers, wrist, and proximal components.

Manual contacts

LEFT HAND

Placed in palm of patient's left hand so that patient may grasp with fingers and thumb and so that wrist may flex toward ulnar side (Fig. 23).

RIGHT HAND

For emphasis of distal joints: Grip with pressure of palmar surface over anterior-ulnar aspect of patient's forearm, so as to control pronation and proximal components of motion.

For emphasis of shoulder and elbow: Pressure of palmar surface on posterior-medial surface of patient's arm, so as to control internal rotation and proximal components of motion (Fig. 23).

For emphasis of scapula: Pressure of palmar surface of hand over anterior-medial axillary space and acromion process.

For mass closing of hand and for emphasis of thumb motion: Grasp patient's left thumb with thumb and index finger of right hand—contact of physical therapist's thumb and index finger should be medially and laterally at interphalangeal joint of patient's thumb, so as to control rotation of patient's thumb as well as flexion and abduction components. Physical therapist's left hand should be placed with palmar surface of hand and fingers on palmar surface of hand and fingers of patient's left hand. Physical therapist's left hand prevents range of motion occurring proximally until fingers flex and adduct to ulnar side. Flexion of fingers and thumb flexion–abduction may be resisted.

Commands

PREPARATORY

"You are to squeeze my hand, turn it, and pull it down and toward your chest, bending your elbow."

ACTION

"Pull!" "Squeeze my hand!" "Turn it!" "Bend your elbow!" "Pull it down to your chest!"

Pattern analysis

SCAPULA

Motion components: Rotation, abduction (medial angle), depression anteriorly (acromion).

Major muscle components: Pectoralis minor, subclavius—acting upon clavicle.

SHOULDER

Motion components: Extension, adduction, internal rotation.

Major muscle components: Subscapularis, pectoralis major—sternal portion.

ELBOW

Motion components: Flexion.

Major muscle components: Biceps brachii—short head, brachialis.

FOREARM

Motion components: Pronation.

Major muscle components: Pronator teres.

WRIST, FINGERS, AND THUMB

See extension–adduction–internal rotation (with elbow straight) pattern.

Range limiting factors

Tension or contracture of any muscles of the flexion–abduction–external rotation (with elbow extension) pattern (Fig. 20).

Thrusting patterns of the upper extremity are primitive in that they are closely related to locomotion in the prone posture when the elbows are fully extended and weight is supported on the hands. These movements are a part of "push-ups." They also promote reaching for an object with extending elbow and with the hand opening in preparation for grasp. As with all patterns, thrusting may be performed in any position or posture which permits the desired range of motion to be performed. Thrusting patterns may be initiated by use of the stretch reflex; the response is a rapid and forceful movement. In the shortened range, approximation may be applied by manual contact at the base of the palm. The command to "Thrust" is always given sharply and is timed with stretch reflex; or, when approximation is used in the shortened range, the command is "Hold."

Thrusting is a variation of the upper extremity patterns. While in the specific patterns of facilitation, opening of the hand is consistent with abduction of the shoulder, in thrusting opening of the hand is consistent with adduction of the shoulder. In the specific patterns all rotation occurs in the same direction; in thrusting the rotation of shoulder and forearm are opposite. The more distal joints move in harmony with the rotation of the forearm. The wrist and elbow always extend. The scapula moves in the direction of the thrust and is, therefore, protracted. The serratus anterior muscle is primarily responsible for the movement of the scapula. The pectoral muscles contribute strongly to the shoulder movement. Thrusting movements may be reversed. During reversal all components of motion are exactly opposite.

Ulnar extensor thrust

The proximal and intermediate components of motion are those of the flexion–adduction–external rotation with elbow extension pattern. The distal components are those of the extension–abduction–internal rotation pattern. In the lengthened range, or starting position, the hand is closed and wrist is flexed toward the radial side; the forearm is supinated; the elbow is completely flexed; and the shoulder is extended in abduction. As the subject thrusts, the hand opens and wrist extends toward the ulnar side, the forearm pronates, the elbow extends, and the shoulder flexes and adducts so that the open hand is reached upward and across the nose and eyes.

Distal manual contacts are essentially the same as those for extension–abduction–internal rotation pattern; that is, the therapist's right hand is placed on the ulnar side of the subject's right hand so as to passively flex the fingers toward the radial side. The therapist grasps the subject's forearm with the left hand near the subject's wrist so as to superimpose stretch and resistance on the elbow extensors and flexors and adductors of the shoulder. The serratus anterior is also stretched by this maneuver. During reversal the therapist allows the subject to grip the fingers and to pull with wrist flexing toward the radial side. As the forearm supinates, the elbow flexes, and the shoulder extends and abducts.

Radial extensor thrust

The proximal and intermediate components are those of the extension–adduction–internal rotation with elbow extension pattern. The distal components are those of the flexion–abduction–external rotation pattern. In the lengthened range, or starting position, the hand is closed toward the ulnar side; the forearm is pronated; the elbow is completely flexed; and the shoulder is flexed in abduction. As the subject thrusts, the hand opens and the wrist extends toward the radial side, the forearm supinates, the elbow extends, and the shoulder extends and adducts so that the open hand is reached downward and across the body toward the opposite hip.

Distal manual contacts are essentially the same as those for the flexion–abduction–external rotation pattern; that is, the therapist's right hand is placed on the radial side of the subject's left hand so as to passively flex the fingers toward the ulnar side. The therapist grasps the subject's forearm with the left hand near the subject's wrist so as to superimpose stretch and resistance on the elbow extensors and extensors and adductors of the shoulder as well as the serratus anterior. During reversal the therapist allows the subject to grip the fingers and to pull with wrist flexing toward the ulnar side as the forearm pronates, the elbow flexes, and the shoulder flexes and abducts.

Thrusting may be performed in any position which permits the desired range of motion. Resisting the thrust may enhance the patient's ability to advance the upper extremities during prone locomotion. One extremity may be exercised as the patient leans on the opposite elbow, or both extremities may participate with alternating reciprocal movements. As the patient reverses the movements, resistance will help him to advance his trunk on the mat.

LOWER EXTREMITY
Flexion–Adduction–External Rotation

Fig. 24. With knee straight.

Antagonistic pattern

Extension–abduction–internal rotation (with knee straight) (Fig. 27).

Components of motion

Toes extend and abduct (medial toes more than lateral) toward tibial side, foot and ankle dorsiflex with inversion, knee remains straight, hip flexes, adducts, and externally rotates.

Normal timing

Action is from distal to proximal, i.e., action occurs first at toes, then foot and ankle, then at hip.

Timing for emphasis

HIP

Allow beginning rotation to occur at toes, foot and ankle, and hip, but do not allow full range of toe extension with abduction, and foot and ankle dorsiflexion and inversion to occur, until hip begins to flex and adduct with external rotation.

Note: If normal timing is prevented by excessive resistance to weaker distal components, action cannot occur proximally. Resist stronger distal components, but guide weaker distal components through their optimal range of motion in accordance with normal timing.

ANKLE AND FOOT

Allow beginning rotation to occur at toes, foot and ankle, and hip, but do not allow full range of toe extension with abduction, and hip flexion–adduction to occur, until foot and ankle begin to dorsiflex and invert.

Note: Resist stronger proximal and distal components, but guide weaker distal components through their optimal range of motion in accordance with normal timing.

TOES

Allow beginning rotation to occur at toes, foot and ankle, and hip, but do not allow full range of foot and ankle dorsiflexion with inversion, and hip flexion–adduction to occur, until toes begin to extend and abduct toward tibial side.

Note: Resist stronger proximal components. Emphasis may be placed on metatarsal-phalangeal joints, or interphalangeal points, or on a specific joint of a single toe.

Manual contacts

1. PATIENT IS ABLE TO WORK THROUGH FULL RANGE
OF PATTERN

Right hand: Pressure of palmar surface of hand on medial aspect of dorsal surface of foot, as far

Individual patterns   55

distal as a firm grip will permit. Avoid pressure on plantar surface of foot (Fig. 24).

Left hand: Pressure of palmar surface of hand or with fingers in close approximation on anterior-medial aspect of thigh proximal to patella (Fig. 24).

2. PATIENT HAS DIFFICULTY IN INITIATING MOTION

Right hand: As in 1.

Left hand: Pressure of palmar surface of hand or with fingers in close approximation on the posterior-medial aspect of thigh proximal to popliteal space; or pressure of fingers in close approximation on medial aspect of right heel.

Commands

PREPARATORY

"You are to turn your heel in, and pull your foot up and across your body."

ACTION

"Pull!" "Pull your foot in and up!" "Pull it up and away from me!"

Pattern analysis

HIP

Motion components: Flexion, adduction, and external rotation.

Major muscle components: Psoas minor, psoas major, iliacus, obturator externus, pectineus, gracilis, adductor brevis, adductor longus, sartorius (hip flexion component), rectus femoris—medial portion (hip flexion component).

KNEE

Straight—no motion.

ANKLE, FOOT, AND TOES

Motion components: Dorsiflexion, inversion of ankle and foot, toe extension with abduction toward tibial side.

Major muscle components: Tibialis anterior, extensor digitorum longus, extensor hallucis longus, extensor digitorum brevis, abductor hallucis, dorsal interossei, lumbricales.

Range limiting factors

Tension or contracture of any muscles of extension–abduction–internal rotation (with knee straight) pattern (Fig. 27).

LOWER EXTREMITY
Flexion–Adduction–External Rotation

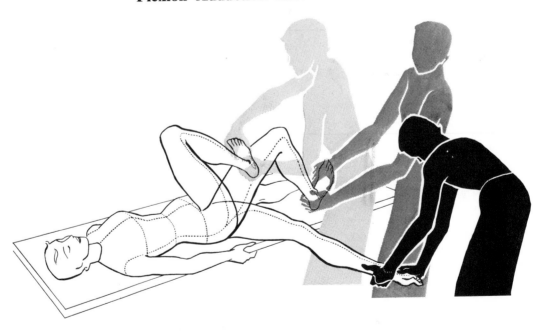

Fig. 25. With knee flexion.

Antagonistic pattern

Extension–abduction–internal rotation (with knee extension) (Fig. 28).

Components of motion

Toes extend and abduct (medial toes more than lateral) toward tibial side, foot and ankle dorsiflex and invert, knee flexes with tibia externally rotating on femur, hip flexes, adducts, and externally rotates.

Normal timing

Action is from distal to proximal, i.e., action occurs first at toes, then foot and ankle, then at knee, then at hip.

Timing for emphasis

HIP

Allow beginning rotation to occur at toes, foot and ankle, knee, and hip, but do not allow full range of toe extension with abduction, foot and ankle dorsiflexion with inversion, and knee flexion to occur, until hip begins to flex and adduct with external rotation.

Note: If normal timing is prevented by excessive resistance to weaker distal components, action cannot occur proximally. Resist stronger distal components, but guide weaker distal components through their optimal range of motion in accordance with normal timing.

KNEE

Allow beginning rotation to occur at toes, foot and ankle, knee, and hip, but do not allow full range of toe extension with abduction, foot and ankle dorsiflexion with inversion, and hip flexion–adduction to occur, until knee begins to flex with external rotation of tibia on femur.

Note: Resist stronger proximal and distal components, but guide weaker distal components through their optimal range of motion in accordance with normal timing.

ANKLE AND FOOT

Allow beginning rotation to occur at toes, foot and ankle, knee, and hip, but do not allow full range of toe extension with abduction, knee flexion, and hip flexion–adduction to occur, until foot and ankle begin to dorsiflex and invert.

Note: Resist stronger proximal and distal components, but guide weaker distal components through their optimal range of motion in accordance with normal timing.

TOES

Allow beginning rotation to occur at toes, foot and ankle, knee, and hip, but do not allow full range of toe extension with abduction, foot and ankle dorsiflexion with inversion, knee flexion, and hip flexion–adduction to occur, until toes begin to extend and abduct toward tibial side.

Note: Resist stronger proximal components. Emphasis may be placed on metatarsal-phalangeal joints or interphalangeal joints or a specific joint of a single toe.

Manual contacts

1. PATIENT IS ABLE TO WORK THROUGH FULL RANGE OF PATTERN

Right hand: Pressure of palmar surface of hand on medial aspect of dorsal surface of foot, as far distal as a firm grip will permit. Avoid pressure on plantar surface of foot (Fig. 25).

Left hand: Pressure of palmar surface of hand or with fingers in close approximation on anterior-medial aspect of thigh proximal to patella; or pressure of fingers in close approximation on medial aspect of heel (Fig. 25).

2. PATIENT HAS DIFFICULTY IN INITIATING MOTION

Right hand: As in 1.

Left hand: Pressure of palmar surface of hand or with fingers in close approximation on the posterior-medial aspect of thigh proximal to popliteal space.

Commands

PREPARATORY

"You are to turn your heel, and pull your foot up and across your body, and bend your knee."

ACTION

"Pull!" "Pull your foot in and up!" "Bend your knee!" "Pull it up and away from me!"

Pattern analysis

HIP

Motion components: Flexion, adduction, and external rotation.

Major muscle components: Psoas minor, psoas major, iliacus, obturator externus, pectineus, gracilis, adductor brevis, adductor longus, sartorius (hip flexion component).

KNEE

Motion components: Flexion with tibia externally rotating on femur.

Major muscle components: Semitendinosus, semimembranosus, sartorius, gracilis (knee flexion component).

ANKLE, FOOT, AND TOES

See flexion–adduction–external rotation (with knee straight) pattern.

Range limiting factors

Tension or contracture of any muscles of extension–abduction–internal rotation (with knee extension) pattern (Fig. 28).

LOWER EXTREMITY
Flexion–Adduction–External Rotation

Fig. 26. With knee extension.

Antagonistic pattern

Extension–abduction–internal rotation (with knee flexion) (Fig. 29).

Components of motion

Toes extend and abduct (medial toes more than lateral) toward tibial side, foot and ankle dorsiflex with inversion, knee extends with tibia externally rotating on femur, hip flexes, adducts, and externally rotates.

Normal timing

Action is from distal to proximal, i.e., action occurs first at toes, then foot and ankle, then at knee, then at hip.

Timing for emphasis
HIP

Allow beginning rotation to occur at toes, foot and ankle, knee, and hip, but do not allow full range of toe extension with abduction, foot and ankle dorsiflexion with inversion, and knee extension to occur, until hip begins to flex and adduct with external rotation.

Note: If normal timing is prevented by excessive resistance to weaker distal components, action cannot occur proximally. Resist stronger distal components, but guide weaker distal components through their optimal range of motion in accordance with normal timing.

KNEE

Allow beginning rotation to occur at toes, foot and ankle, knee, and hip, but do not allow full range of toe extension with abduction, foot and ankle dorsiflexion with inversion, and hip flexion–adduction to occur until knee begins to extend with external rotation of tibia on femur.

Note: Resist stronger proximal and distal components, but guide weaker distal components through their optimal range of motion in accordance with normal timing.

ANKLE AND FOOT

Allow beginning rotation to occur at toes, foot and ankle, knee, and hip, but do not allow full range of toe extension with abduction, knee extension, and hip flexion–adduction to occur, until foot and ankle begin to dorsiflex and invert.

Note: Resist stronger proximal and distal components, but guide weaker distal components through their optimal range of motion in accordance with normal timing.

TOES

Allow beginning rotation to occur at toes, foot and ankle, knee, and hip, but do not allow full range of foot and ankle dorsiflexion with inversion, knee extension, and hip-flexion–adduction to occur, until toes begin to extend and abduct toward tibial side.

Note: Resist stronger proximal components. Emphasis may be placed on metatarsal-phalangeal joints, or interphalangeal joints, or a specific joint of a single toe.

Manual contacts

1. PATIENT IS ABLE TO WORK THROUGH FULL RANGE OF PATTERN

Right hand: Pressure of palmar surface of hand on medial aspect of dorsal surface of foot, as far distal as a firm grip will permit. Avoid pressure on plantar surface of foot (Fig. 26).

Left Hand: Pressure of palmar surface of hand, or with fingers in close approximation, on anterior-medial aspect of thigh proximal to patella (Fig. 26).

2. PATIENT HAS DIFFICULTY IN INITIATING MOTION

Right hand: As in 1. Fingers may be used to grip foot in order to guide motion.

Left hand: Pressure of palmar surface of hand or with fingers in close approximation on posterior-medial surface of thigh proximal to popliteal space; or pressure of fingers in close approximation, on medial aspect of heel.

Commands

PREPARATORY

"You are to kick your foot up and in across your body, and straighten your knee."

ACTION

"Kick!"

Pattern analysis

HIP

Motion components: Flexion, adduction, external rotation.

Major muscle components: Psoas minor, psoas major, iliacus, obturator externus, pectineus, gracilis, adductor brevis, adductor longus, rectus femoris-medial portion (hip flexion component).

KNEE

Motion components: Extension with tibia externally rotating on femur.

Major muscle components: Rectus femoris—medial portion, vastus medialis, articularis genu.

ANKLE, FOOT, AND TOES

See flexion–adduction–external rotation (with knee straight) pattern.

Range limiting factors

Tension or contracture of any muscles of extension–abduction–internal rotation (with knee flexion) pattern (Fig. 29).

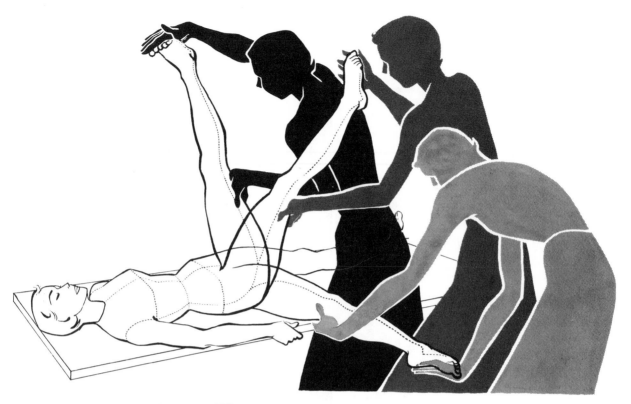

Fig. 27. With knee straight.

Antagonistic pattern

Flexion–adduction–external rotation (with knee straight) (Fig. 24).

Components of motion

Toes flex and adduct (lateral toes more than medial) toward fibular side, foot and ankle plantar flex with eversion, knee remains straight, hip extends, abducts, and internally rotates.

Normal timing

Action is from distal to proximal, i.e., action occurs first at toes, then foot and ankle, then at hip.

Timing for emphasis

HIP

Allow beginning rotation to occur at toes, foot and ankle, and hip, but do not allow full range of toe flexion with adduction, and foot and ankle plantar flexion with eversion to occur, until hip begins to extend and abduct with internal rotation.

Note: If normal timing is prevented by excessive resistance to weaker distal components, action cannot occur proximally. Resist stronger distal components, but guide weaker distal components through their optimal range of motion in accordance with normal timing.

ANKLE AND FOOT

Allow beginning rotation to occur at toes, foot, and ankle, and hip, but do not allow full range of toe flexion with adduction, and hip extension–abduction to occur, until foot and ankle begin to plantar flex and evert.

Note: Resist stronger proximal and distal components, but guide weaker distal components through their optimal range of motion in accordance with normal timing.

TOES

Allow beginning rotation to occur at toes, foot and ankle, and hip, but do not allow full range of foot and ankle plantar flexion with eversion to occur, until toes begin to flex and adduct toward fibular side.

Note: Resist stronger proximal components. Emphasis may be placed on metatarsal-phalangeal joints, or interphalangeal joints, or on a specific joint of a single toe.

Manual contacts

1. PATIENT IS ABLE TO WORK THROUGH FULL RANGE OF PATTERN

Right hand: Pressure of palmar surface of hand and fingers on lateral aspect of plantar surface of foot and toes (Fig. 27).

Left hand: Pressure of palmar surface or with fingers in close approximation on posterior-lateral aspect of thigh proximal to popliteal space (Fig. 27).

2. PATIENT HAS DIFFICULTY IN INITIATING MOTION

Right hand: As in 1.

Left hand: As in 1, or with pressure applied by palmar surface with fingers free of contact on lateral aspect of heel.

Commands

PREPARATORY

"You are to turn your heel, and push your foot down and out toward me."

ACTION

"Push!" "Push your foot down and out!" "Keep your knee straight!" "Push down at the hip, toward me!"

Pattern analysis

HIP

Motion components: Extension, abduction, internal rotation.

Major muscle components: Gluteus medius, gluteus minimus, biceps femoris (hip extension component).

KNEE

Straight—no motion.

ANKLE, FOOT, AND TOES

Motion components: Plantar flexion, eversion of ankle and foot, flexion with adduction of toes toward fibular side.

Major muscle components: Gastrocnemius—lateral head, soleus—lateral portion, peroneus longus, flexor digitorum longus, flexor digitorum brevis, flexor hallucis brevis, adductor hallucis, flexor digiti quinti brevis, quadratus plantae, plantar interossei, lumbricales.

Range limiting factors

Tension or contracture of any muscles of flexion–adduction–external rotation (with knee straight) pattern (Fig. 24).

LOWER EXTREMITY
Extension–Abduction–Internal Rotation

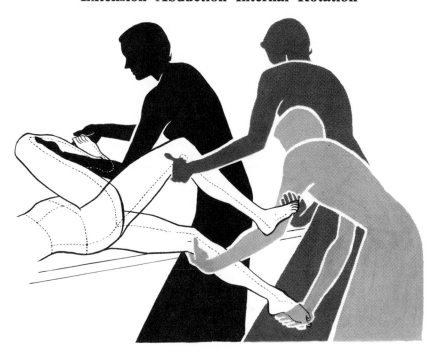

Fig. 28. With knee extension.

Antagonistic pattern

Flexion–adduction–external rotation (with knee flexion) (Fig. 25).

Components of motion

Toes flex and adduct (lateral toes more than medial) toward fibular side, foot and ankle plantar flex with eversion, knee extends with tibia internally rotating on femur, hip extends, abducts, and internally rotates.

Normal timing

Action is from distal to proximal, i.e., action occurs first at toes, then foot and ankle, then knee, then at hip.

Timing for emphasis

HIP

Allow beginning rotation to occur at toes, foot and ankle, knee, and hip, but do not allow full range of flexion with adduction of toes, foot and ankle plantar flexion with eversion, and knee extension to occur, until hip begins to extend and abduct with internal rotation.

Note: If normal timing is prevented by excessive resistance to weaker distal components, action cannot occur proximally. Resist stronger distal components, but guide weaker distal components through their optimal range of motion in accordance with normal timing.

KNEE

Allow beginning rotation to occur at toes, foot and ankle, knee, and hip, but do not allow full range of toe flexion with adduction, foot and ankle plantar flexion with eversion, and hip extension–abduction to occur, until knee begins to extend with internal rotation of tibia on femur.

Note: Resist stronger proximal and distal components, but guide weaker distal components through their optimal range of motion in accordance with normal timing.

ANKLE AND FOOT

Allow beginning rotation to occur at toes, foot and ankle, knee, and hip, but do not allow full range of toe flexion with adduction, knee extension, and hip extension–abduction to occur, until foot and ankle begin to plantar flex and evert.

Note: Resist stronger proximal and distal components, but guide weaker distal components through their optimal range of motion in accordance with normal timing

TOES

Allow beginning rotation to occur at toes, foot and ankle, knee, and hip, but do not allow full range of foot and ankle plantar flexion with eversion, knee extension, and hip extension–abduction to occur, until toes begin to flex and adduct toward fibular side.

Note: Resist stronger proximal components. Emphasis may be placed on metatarsal-phalangeal joints, or interphalangeal joints, or on a specific joint of a single toe.

Manual contacts

1. PATIENT IS ABLE TO WORK THROUGH FULL RANGE OF PATTERN

Right hand: Pressure of palmar surface of hand and fingers on lateral aspect of plantar surface of toes and foot (Fig. 28).

Left hand: Pressure of palmar surface or with fingers in close approximation on posterior-lateral aspect of thigh proximal to popliteal space (Fig. 28).

2. PATIENT HAS DIFFICULTY IN INITIATING MOTION

Right hand: As in 1.

Left hand: As in 1, or with pressure applied by palmar surface with fingers free of contact on lateral aspect of heel.

Commands

PREPARATORY

"You are to turn your heel, and push your foot down and out toward me, and straighten your knee."

ACTION

"Push!" "Push your foot down and out!" "Push down at the hip and knee, toward me!"

Pattern analysis

HIP

Motion components: Extension, abduction, internal rotation.

Major muscle components: Gluteus medius, gluteus minimus.

KNEE

Motion components: Extension with tibia internally rotating on femur.

Major muscle components: Vastus intermedius, vastus lateralis, articularis genu.

ANKLE, FOOT, AND TOES

See extension–abduction–internal rotation (with knee straight) pattern.

Range limiting factors

Tension or contracture of any muscles of flexion–adduction–external rotation (with knee flexion) pattern (Fig. 25).

LOWER EXTREMITY
Extension–Abduction–Internal Rotation

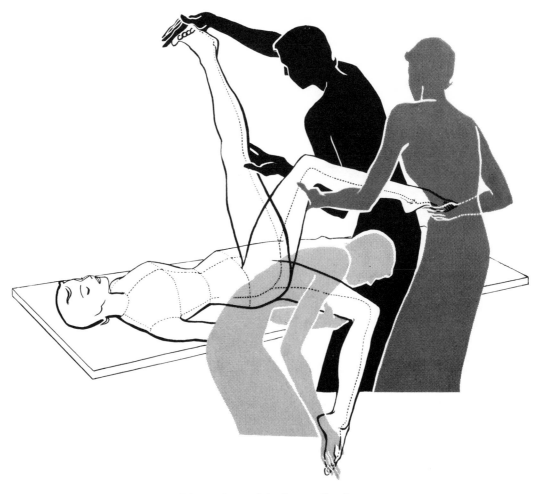

Fig. 29. With knee flexion.

Antagonistic pattern

Flexion–adduction–external rotation (with knee extension) (Fig. 26).

Components of motion

Toes flex and adduct (lateral toes more than medial) toward fibular side, foot and ankle plantar flex with eversion, knee flexes with tibia internally rotating on femur, hip extends, abducts, and internally rotates.

Normal timing

Action is from distal to proximal, i.e., action occurs first at toes, then foot and ankle, then at knee, then at hip.

Timing for emphasis

HIP

Allow beginning rotation to occur at toes, foot and ankle, knee, and hip, but do not allow full range of toe flexion with adduction, foot and ankle plantar flexion with eversion, and knee flexion to occur, until hip begins to extend and abduct with internal rotation.

Note: If normal timing is prevented by excessive resistance to weaker distal components, action cannot occur proximally. Resist stronger distal components, but guide weaker distal components through their optimal range of motion in accordance with normal timing.

KNEE

Allow beginning rotation to occur at toes, foot and ankle, knee, and hip, but do not allow full range of toe flexion with adduction, foot and ankle plantar flexion with eversion, and hip extension–abduction to occur, until knee begins to flex with internal rotation of tibia on femur.

Note: Resist stronger proximal and distal components, but guide weaker distal components through their optimal range of motion in accordance with normal timing.

Patterns of facilitation

Allow beginning rotation to occur at toes, foot and ankle, knee, and hip, but do not allow full range of toe flexion with adduction, knee flexion, and hip extension–abduction to occur, until foot and ankle begin to plantar flex and evert.

Note: Resist stronger proximal and distal components, but guide weaker distal components through their optimal range of motion in accordance with normal timing.

TOES

Allow beginning rotation to occur at toes, foot and ankle, knee, and hip, but do not allow full range of foot and ankle plantar flexion with eversion, knee flexion, hip extension–abduction to occur, until toes begin to flex and adduct toward fibular side.

Note: Resist stronger proximal components. Emphasis may be placed on metatarsal-phalangeal joints, or interphalangeal joints, or a specific joint of a single toe.

Manual contacts

1. PATIENT IS ABLE TO WORK THROUGH FULL RANGE OF PATTERN

Right hand: Pressure of palmar surface of hand and fingers on lateral aspect of plantar surface of foot and toes (Fig. 29).

Left hand: Pressure of palmar surface of hand or with fingers in close approximation on posterior-lateral aspect of thigh proximal to popliteal space (Fig. 29).

2. PATIENT HAS DIFFICULTY IN INITIATING MOTION

Right hand: Same as 1.
Left hand: Same as 1.

Commands

PREPARATORY

"You are to turn your heel, and push your foot down and out toward me, and bend your knee."

ACTION

"Push!" "Push your toes down and out!" "Bend your knee!" "Push down at the hip, toward me!"

Pattern analysis

HIP

Motion components: Extension, abduction, internal rotation.

Major muscle components: Gluteus medius, gluteus minimus, biceps femoris (hip extension component).

KNEE

Motion components: Flexion with internal rotation of tibia on femur.

Major muscle components: Biceps femoris, popliteus, gastrocnemius—lateral head.

ANKLE, FOOT, AND TOES

See extension–abduction–internal rotation (with knee straight) pattern.

Range limiting factors

Tension or contracture of any muscles of flexion–adduction–external rotation (with knee extension) pattern (Fig. 26).

LOWER EXTREMITY
Flexion–Abduction–Internal Rotation

Fig. 30. With knee straight.

Antagonistic pattern

Extension–adduction–external rotation (with knee straight) (Fig. 33).

Components of motion

Toes extend and abduct (lateral toes more than medial) toward the fibular side, foot and ankle dorsiflex with eversion, knee remains straight, hip flexes, abducts, and internally rotates.

Normal timing

Action is from distal to proximal, i.e., action occurs first at toes, then foot and ankle, then at hip.

Timing for emphasis

HIP

Allow beginning rotation to occur at toes, foot and ankle, but do not allow full range of toe extension with abduction, and foot and ankle dorsiflexion with eversion to occur, until hip begins to flex and abduct with internal rotation.

Note: If normal timing is prevented by excessive resistance to weaker distal components, action cannot occur proximally. Resist stronger distal components, but guide weaker distal components through their optimal range of motion in accordance with normal timing.

ANKLE AND FOOT

Allow beginning rotation to occur at toes, foot and ankle, and hip, but do not allow full range of toe extension with abduction and hip flexion–abduction to occur, until foot and ankle begin to dorsiflex and evert.

Note: Resist stronger proximal and distal components, but guide weaker distal components through their optimal range of motion in accordance with normal timing.

TOES

Allow beginning rotation to occur at toes, foot and ankle, and hip, but do not allow full range of foot and ankle dorsiflexion with eversion and hip flexion–abduction to occur, until toes begin to extend and abduct toward fibular side.

Note: Resist stronger proximal components. Emphasis may be placed on metatarsal-phalangeal joints, or interphalangeal joints, or on a specific joint of a single toe.

Manual contacts

1. PATIENT IS ABLE TO WORK THROUGH FULL RANGE OF PATTERN

Left hand: Pressure of palmar surface of hand on lateral aspect of dorsal surface of foot as far distal as a firm grip will permit. Avoid pressure on plantar surface of foot (Fig. 30).

Right hand: Pressure of palmar surface of hand or with fingers in close approximation on anterior-lateral aspect of thigh proximal to patella (Fig. 30).

2. PATIENT HAS DIFFICULTY IN INITIATING MOTION

Left hand: As in 1.

Right hand: Pressure of palmar surface of hand or with fingers in close approximation on posterior-lateral aspect of thigh proximal to popliteal space; or pressure of palmar surface of hand on lateral aspect of heel.

Commands

PREPARATORY

"You are to turn your heel, and pull your foot up and out as far as possible."

ACTION

"Pull!" "Pull your foot up and out!" "Lift it up toward me!"

Pattern analysis

HIP

Motion components: Flexion, abduction, internal rotation.

Major muscle components: Tensor fasciae latae, rectus femoris—lateral portion (hip flexion component).

KNEE

Straight—no motion.

ANKLE, FOOT, AND TOES

Motion components: Dorsiflexion, eversion of ankle and foot, toe extension with abduction toward fibular side.

Major muscle components: Extensor digitorum longus, extensor hallucis longus, peroneus brevis, peroneus tertius, extensor digitorum brevis, abductor digiti quinti, dorsal interossei, lumbricales.

Range limiting factors

Tension or contracture of any muscles of extension–adduction–external rotation (with knee straight) pattern (Fig. 33).

LOWER EXTREMITY
Flexion–Abduction–Internal Rotation

Fig. 31. With knee flexion.

Antagonistic pattern

Extension–adduction–external rotation (with knee extension) (Fig. 34).

Components of motion

Toes extend and abduct (lateral toes more than medial) toward fibular side, foot and ankle dorsiflex with eversion, knee flexes with tibia internally rotating on femur, hip flexes, abducts, and internally rotates.

Normal timing

Action is from distal to proximal, i.e., action occurs first at toes, then foot and ankle, then knee, then at hip.

Timing for emphasis

HIP

Allow beginning rotation to occur at toes, foot and ankle, knee, and hip, but do not allow full range of toe extension with abduction, foot and ankle dosriflexion with eversion, and knee flexion to occur, until hip begins to flex and abduct with internal rotation.

Note: If normal timing is prevented by excessive resistance to weaker distal components, action cannot occur proximally. Resist stronger distal components, but guide weaker distal components through their optimal range of motion in accordance with normal timing.

KNEE

Allow beginning rotation to occur at toes, foot and ankle, and knee, and hip, but do not allow full range of toe extension with abduction, foot and ankle dorsiflexion with eversion, and hip flexion–abduction to occur, until knee begins to flex with internal rotation of tibia on femur.

Note: Resist stronger proximal and distal components, but guide weaker distal components through their optimal range of motion in accordance with normal timing.

ANKLE AND FOOT

Allow beginning rotation to occur at toes, foot and ankle, knee, and hip, but do not allow full range of toe extension with abduction, knee flexion and hip flexion–abduction to occur, until foot and ankle begin to dorsiflex and evert.

Note: Resist stronger proximal and distal components, but guide weaker distal components through their optimal range of motion in accordance with normal timing.

TOES

Allow beginning rotation to occur at toes, foot and ankle, knee, and hip, but do not allow full range of foot and ankle dorsiflexion with eversion, knee flexion, hip flexion–abduction to occur, until toes begin to extend and abduct toward fibular side.

Note: Resist stronger proximal components. Emphasis may be placed on metatarsal-phalangeal joints, or interphalangeal joints, or a specific joint of a single toe.

Manual contacts

1. PATIENT IS ABLE TO WORK THROUGH FULL RANGE OF PATTERN

Left hand: Pressure of palmar surface of hand on lateral aspect of dorsal surface of foot as far distal as a firm grip will permit. Avoid pressure on plantar surface of foot (Fig. 31).

Right hand: Pressure of palmar surface of hand or with four fingers in close approximation on anterior-lateral aspect of thigh proximal to patella (*illustrated—middle and shortened ranges*).

2. PATIENT HAS DIFFICULTY IN INITIATING MOTION

Left hand: As in 1.

Right hand: Pressure of palmar surface of hand on posterior-lateral surface of thigh proximal to popliteal space, or pressure of palmar surface on lateral aspect of heel (Fig. 31—lengthened range).

Commands

PREPARATORY

"You are to turn your heel, and pull your foot up and out, and bend your knee."

ACTION

"Pull!" "Pull your foot up and out!" "Bend your knee!" "Pull up at the hip—toward me!"

Pattern analysis

HIP

Motion components: Flexion, abduction, internal rotation.
Major muscle components: Tensor fasciae latae.

KNEE

Motion components: Flexion with tibia internally rotating on femur.
Major muscle components: Biceps femoris, popliteus.

ANKLE, FOOT, AND TOES

See flexion–abduction–internal rotation (with knee straight) pattern.

Range limiting factors

Tension or contracture of any muscles of extension–adduction–external rotation (with knee extension) pattern (Fig. 34).

LOWER EXTREMITY
Flexion–Abduction–Internal Rotation

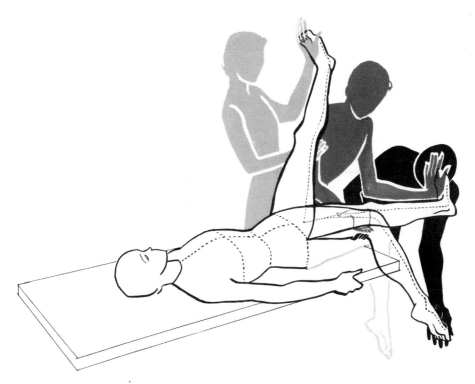

Fig. 32. With knee extension

Antagonistic pattern

Extension–adduction–external rotation (with knee flexion) (Fig. 35).

Components of motion

Toes extend and abduct (lateral toes more than medial) toward fibular side, foot and ankle dorsiflex with eversion, knee extends with tibia internally rotating on femur, hip flexes, abducts, and internally rotates.

Normal timing

Action is from distal to proximal, i.e., action occurs at toes, then foot and ankle, then knee, and then at hip.

Timing for emphasis

HIP

Allow beginning rotation to occur at toes, foot and ankle, knee, and hip, but do not allow full range of toe extension with abduction, foot and ankle dorsiflexion with eversion, and extension of knee to occur, until hip begins to flex and abduct with internal rotation.

Note: If normal timing is prevented by excessive resistance to weaker distal components, action cannot occur proximally. Resist stronger distal components, but guide weaker distal components through their optimal range of motion in accordance with normal timing.

KNEE

Allow beginning rotation to occur at toes, foot and ankle, knee, and hip, but do not allow full range of toe extension with abduction, foot and ankle dorsiflexion with eversion, and hip flexion–abduction to occur, until knee begins to extend with internal rotation of tibia on femur.

Note: Resist stronger proximal and distal components, but guide weaker distal components through their optimal range of motion in accordance with normal timing.

TOES

Allow beginning rotation to occur at toes, foot and ankle, knee, and hip, but do not allow full range of foot and ankle dorsiflexion with inversion, knee extension, and hip flexion–abduction to occur, until toes begin to extend and abduct toward fibular side.

Note: Resist stronger proximal components. Emphasis may be placed on metatarsal-phalangeal joints, or interphalangeal joints, or on a specific joint of a single toe.

Manual contacts

1. PATIENT IS ABLE TO WORK THROUGH FULL RANGE OF PATTERN

Left hand: Pressure of palmar surface of hand on lateral aspect of dorsal surface of foot as far distal as a firm grip will permit. Avoid pressure on plantar surface of foot (Fig. 32).

Right hand: Pressure of palmar surface of hand or with fingers in close approximation on anterior-lateral surface of thigh proximal to patella.

2. PATIENT HAS DIFFICULTY IN INITIATING MOTION

Left hand: As in 1. Fingers may be used to grip foot in order to guide motion.

Right hand: Pressure of palmar surface of hand or with fingers in close approximation on posterior-lateral aspect of thigh proximal to popliteal space (Fig. 32).

Commands

PREPARATORY

"You are to turn your heel, and kick your foot up and out, and straighten your knee."

ACTION

"Kick!" "Pull your foot up and out!" "Kick it up here!" "Pull up at the hip toward me!"

Pattern analysis

HIP

Motion components: Flexion, abduction, internal rotation.
Major muscle components: Tensor fasciae latae, rectus femoris—lateral portion (hip flexion component).

KNEE

Motion components: Extension with tibia internally rotating on femur.
Major muscle components: Vastus intermedius, vastus lateralis, rectus femoris—lateral portion, articularis genu.

ANKLE, FOOT, AND TOES

See flexion–abduction–internal rotation (with knee straight) pattern.

Range limiting factors

Tension or contracture of any muscle of extension–adduction–external rotation (with knee flexion) pattern (Fig. 35).

LOWER EXTREMITY
Extension–Adduction–External Rotation

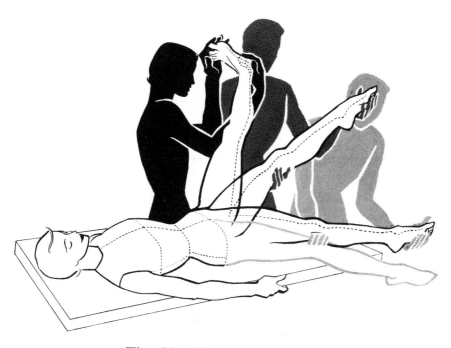

Fig. 33. With knee straight.

Antagonistic pattern

Flexion–abduction–internal rotation (with knee straight) (Fig. 30).

Components of motion

Toes flex and adduct (medial toes more than lateral) toward tibial side, foot and ankle plantar flex with inversion, knee remains straight, hip extends, adducts, and externally rotates.

Normal timing

Action is from distal to proximal, i.e., action occurs first at toes, then foot and ankle, then at hip.

Timing for emphasis

HIP

Allow beginning rotation to occur at toes, foot and ankle, and hip, but do not allow full range of toe flexion with adduction, and foot and ankle plantar flexion with inversion to occur, until hip begins to extend and adduct with external rotation.

Note: If normal timing is prevented by excessive resistance to weaker distal components, action cannot occur proximally. Resist stronger distal components, but guide weaker distal components through their optimal range of motion in accordance with normal timing.

ANKLE AND FOOT

Allow beginning rotation to occur at toes, foot and ankle, and hip, but do not allow full range of toe flexion with adduction, and extension–adduction of hip to occur, until the foot and ankle begin to plantar flex and invert.

Note: Resist stronger proximal and distal components, but guide weaker distal components through their optimal range of motion in accordance with normal timing.

TOES

Allow beginning rotation to occur at toes, foot and ankle, and hip, but do not allow full range of foot and ankle plantar flexion with inversion, and hip extension–adduction to occur, until toes begin to flex and adduct toward tibial side.

Note: Resist stronger proximal components. Emphasis may be placed on metatarsal-phalangeal joints, or interphalangeal joints, or on a specific joint of a single toe.

Manual contacts

1. PATIENT IS ABLE TO WORK THROUGH FULL RANGE OF PATTERN

Left hand: Pressure of palmar surface of hand and fingers on medial aspect of plantar surface of toes and foot (Fig. 33).

Right hand: Pressure of palmar surface of hand on posterior-medial aspect of thigh proximal to popliteal space (Fig. 33 middle and shortened ranges).

2. PATIENT HAS DIFFICULTY IN INITIATING MOTION

Left hand: As in 1.

Right hand: As in 1, or pressure of fingers in close approximation on the medial aspect of heel (Fig. 33—lengthened range).

Commands

PREPARATORY

"You are to turn your heel, and push your foot down and in, away from me."

ACTION

"Push!" "Push your foot down and in!" "Push down at the hip, away from me!"

Pattern analysis

HIP

Motion components: Extension, adduction, external rotation.

Major muscle components: Gluteus maximus, piriformis, gemellus superior, gemellus inferior, obturator internus, quadratus femoris, adductor magnus, semimembranosus and semitendinosus (hip extension components).

KNEE

Straight—no motion.

ANKLE, FOOT, AND TOES

Motion components: Plantar flexion and inversion of ankle and foot, flexion with adduction of toes toward tibial side.

Major muscle components: Plantaris, gastrocnemius—medial head, soleus—medial portion, tibialis posterior, flexor digitorum longus, flexor hallucis longus, quadratus plantae, flexor digitorum brevis, flexor hallucis brevis, plantar interossei, lumbricales.

Range limiting factors

Tension or contracture of any muscles of flexion–abduction–internal rotation (with knee straight) pattern (Fig. 30).

LOWER EXTREMITY
Extension–Adduction–External Rotation

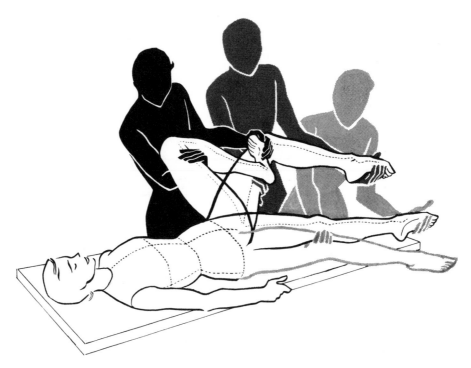

Fig. 34. With knee extension.

Antagonistic pattern

Flexion–abduction–internal rotation (with knee flexion) (Fig. 31).

Components of motion

Toes flex and adduct (medial toes more than lateral) toward tibial side, foot and ankle plantar flex with inversion, knee extends with tibia externally rotating on femur, hip extends, adducts, and externally rotates.

Normal timing

Action is from distal to proximal, i.e., action occurs first at toes, then foot and ankle, then at knee, and then at hip.

Timing for emphasis

HIP

Allow beginning rotation to occur at toes, foot and ankle, knee, and hip, but do not allow full range of toe flexion with adduction, foot and ankle plantar flexion with inversion, and knee extension to occur, until hip begins to extend and adduct with external rotation.

Note: If normal timing is prevented by excessive resistance to weaker distal components, action cannot occur proximally. Resist stronger distal components, but guide weaker distal components through their optimal range of motion in accordance with normal timing.

KNEE

Allow beginning rotation to occur at toes, foot and ankle, knee, and hip, but do not allow full range of toe flexion with adduction, foot and ankle plantar flexion with inversion, and hip extension–adduction to occur, until knee begins to extend with external rotation of tibia on femur.

Note: Resist stronger proximal and distal components, but guide weaker distal components through their optimal range of motion in accordance with normal timing.

ANKLE AND FOOT

Allow beginning rotation to occur at toes, foot, and ankle, knee, and hip, but do not allow full range of toe flexion with adduction, knee extension, and hip extension–adduction to occur, until foot and ankle begin to plantar flex and invert.

Note: Resist stronger proximal and distal components, but guide weaker distal components through their optimal range of motion in accordance with normal timing.

TOES

Allow beginning rotation to occur at toes, foot and ankle, knee, and hip, but do not allow full range of foot and ankle plantar flexion with inversion, knee extension, and hip extension–adduction to occur, until toes begin to flex and adduct toward tibial side.

Note: Resist stronger proximal components. Emphasis may be placed on metatarsal-phalangeal joints, or interphalangeal joints, or on a specific joint of a single toe.

Manual contacts

1. PATIENT IS ABLE TO WORK THROUGH FULL RANGE OF PATTERN

Left hand: Pressure of palmar surface of hand or with fingers in close approximation on medial aspect of plantar surface of toes and foot (Fig. 34).

Right hand: Pressure of palmar surface of hand or with fingers in close approximation on posterior-medial aspect of thigh proximal to popliteal space (Fig. 34).

2. PATIENT HAS DIFFICULTY IN INITIATING MOTION

Left hand: As in 1.
Right hand: As in 1.

Commands

PREPARATORY

"You are to turn your heel, and push your foot down and in, and straighten your knee."

ACTION

"Push!" "Push your foot down and in!" "Push down at the knee and hip, away from me!"

Pattern analysis

HIP

Motion components: Extension, adduction, external rotation.

Major muscle components: Gluteus maximus, piriformis, gemellus superior, gemellus inferior, obturator internus, quadratus femoris, adductor magnus.

KNEE

Motion components: Extension with tibia externally rotating on femur.

Major muscle components: Vastus medialis, articularis genu.

ANKLE, FOOT, AND TOES

See extension–adduction–external rotation (with knee straight) pattern.

Range limiting factors

Tension or contracture of any muscles of flexion–abduction–internal rotation (with knee flexion) pattern (Fig. 31).

LOWER EXTREMITY
Extension–Adduction–External Rotation

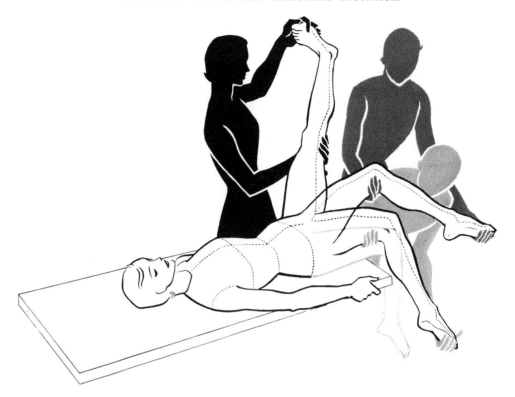

Fig. 35. With knee flexion.

Antagonistic pattern

Flexion–abduction–internal rotation (with knee extension) (Fig. 32).

Components of motion

Toes flex and adduct (medial toes more than lateral) toward tibial side, foot and ankle plantar flex with inversion, knee flexes with tibia externally rotating on femur, hip extends, adducts, and externally rotates.

Normal timing

Action is from distal to proximal, i.e., action occurs first at toes, then foot and ankle, then at knee, and then at hip.

Timing for emphasis

HIP

Allow beginning rotation to occur at toes, foot and ankle, and knee, and hip, but do not allow full range of toe flexion with adduction, foot and ankle plantar flexion with inversion, and knee flexion to occur, until hip begins to extend and adduct with external rotation.

Note: If normal timing is prevented by excessive resistance to weaker distal components, action cannot occur proximally. Resist stronger distal components, but guide weaker distal components through their optimal range of motion in accordance with normal timing.

KNEE

Allow beginning rotation to occur at toes, foot and ankle, knee and hip, but do not allow full range of toe flexion with adduction, foot and ankle plantar flexion with inversion, and hip extension–adduction to occur, until knee begins to flex with external rotation of tibia on femur.

Note: Resist stronger proximal and distal components, but guide weaker distal components through their optimal range of motion in accordance with normal timing.

ANKLE AND FOOT

Allow beginning rotation to occur at toes, foot and ankle, knee, and hip, but do not allow full range of toe flexion with adduction, knee flexion, and hip extension–adduction to occur until foot and ankle begin to plantar flex and invert.

Note: Resist stronger proximal and distal components, but guide weaker distal components through their optimal range of motion in accordance with normal timing.

Allow beginning rotation to occur at toes, foot and ankle, knee, and hip, but do not allow full range of foot and ankle plantar flexion with inversion, knee flexion, and hip extension–adduction to occur, until the toes begin to flex and adduct toward tibial side.

Note: Resist stronger proximal components. Emphasis may be placed on metatarsal-phalangeal joints, or interphalangeal joints, or on a specific joint of a single toe.

Manual contacts

1. PATIENT IS ABLE TO WORK THROUGH FULL RANGE OF PATTERN

Left hand: Pressure of palmar surface of hand or with fingers in close approximation on medial aspect of plantar surface of toes and foot (Fig. 35).

Right hand: Pressure of palmar surface of hand or with fingers in close approximation on posterior-medial aspect of thigh proximal to popliteal space (Fig. 35).

2. PATIENT HAS DIFFICULTY IN INITIATING MOTION

Left hand: As in 1, or with pressure on posterior-medial aspect of heel.
Right hand: As in 1.

Commands

PREPARATORY

"You are to turn your heel, and push your foot down and in, and bend your knee."

ACTION

"Push!" "Push your foot down and in!" "Push down at the hip, away from me!"

Pattern analysis

HIP

Motion components: Extension, adduction, external rotation.

Major muscle components: Gluteus maximus, piriformis, gemellus superior, gemellus inferior, obturator internus, quadratus femoris, adductor magnus, semimembranosus, semitendinosus (hip extension components).

KNEE

Motion components: Flexion with external rotation of tibia on femur.

Major muscle components: Semimembranosus, semitendinosus, gastrocnemius—medial head, plantaris.

ANKLE, FOOT, AND TOES

See extension–adduction–external rotation (with knee straight) pattern.

Range limiting factors

Tension or contracture of any muscles of flexion–abduction—internal rotation (with knee extension) pattern (Fig. 32).

Timing for emphasis and range of motion variation

Timing for emphasis of various pivots of action has been presented for each pattern of facilitation. When timing for emphasis is applied as a technique, the range of motion varies according to the pivot emphasized. When a proximal pivot is emphasized, complete range of motion of that pivot is desired. When a distal pivot is emphasized, range of motion of the proximal pivot may be prevented by resist-ance as a means for stimulating or increasing the responses of the distal pivot. The figures on this and the facing page portray examples of variations in range of motion. The components of motion and the major muscle components are the same as when the pattern is performed with complete range of all pivots of action.

UPPER EXTREMITY
Flexion–Abduction–External Rotation

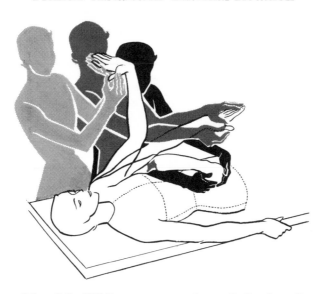

Fig. 36. With mass opening of the hand.

Emphasis of the distal pivots, fingers, and hand prevents complete range of shoulder motion. For complete range of motion see the flexion–abduction–external rotation (with elbow straight) pattern (Fig. 18).

LOWER EXTREMITY
Flexion–Adduction–External Rotation

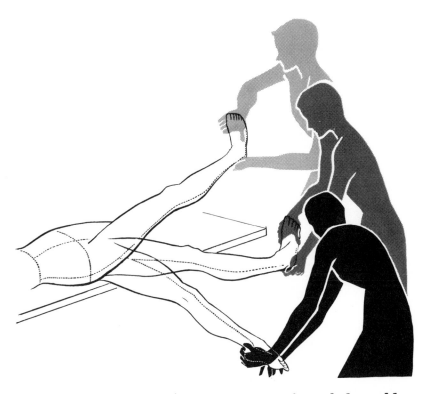

Fig. 37. With dorsiflexion and inversion of the ankle.

Emphasis of the distal pivots, the foot and ankle, prevents complete range of hip motion. For complete range of motion see the flexion–adduction–external rotation (with knee straight) pattern (Fig.24).

Timing for emphasis and range of motion variation

2. Techniques for facilitation

Introduction

Within the repertoire of techniques which are superimposed upon movement and posture, there are certain procedures that are considered basic to the approach. These basic procedures become a part of the treatment of every patient insofar as his medical condition permits their use. In the broadest sense, the basic procedures may be used with or without the patient's complete cooperation; if the patient moves, the physical therapist's maneuvers guide and influence the patient's response. The procedures have to do with how the therapist approaches the patient, how manual contacts are made effective, how the therapist communicates with the patient, how the therapist opposes the patient's effort and at the same time becomes a part of his effort, how coordination is brought about through timing, and how reinforcement is used to increase response and to circumvent fatigue. Thus, the basic procedures include: Manual Contacts, Commands and Communication, Stretch, Traction and Approximation, Maximal Resistance, Normal Timing, and Reinforcement which includes Timing for Emphasis, Combining Patterns, and Recuperative Motion.

Beyond the basic procedures there is a battery of specific techniques which are, for the most part, dependent upon the patient's cooperation and his voluntary effort. Whenever and wherever possible the patient's voluntary effort is used to promote volitional control of movement and posture. The coupling of the patient's voluntary effort with resistance, graded appropriately by the physical therapist, permits the use of specific techniques for stimulation or facilitation and for relaxation or inhibition. Facilitation and inhibition are inseparable: A technique which promotes response or facilitation of the agonist simultaneously promotes relaxation or inhibition of the antagonist.

The specific techniques may be analyzed as to whether they are directed primarily to the agonist; whether the antagonist is used primarily to facilitate the agonist; or whether the antagonist is used primarily for relaxation or inhibition of the antagonist itself. Again, because facilitation and inhibition are inseparable, there is an overlapping of effect.

Techniques directed primarily to the agonist are Repeated Contractions and Rhythmic Initiation; repetition of movement in the desired direction is used. Techniques which employ the antagonist for facilitation of the agonist are the Reversal of Antagonists techniques, Slow Reversal, Slow Reversal-Hold, and Rhythmic Stabilization. When the antagonist is used to facilitate the agonist, an active interaction between agonist and antagonist is demanded. The techniques which are directed primarily toward relaxation or inhibition of the antagonist are termed Relaxation Techniques and include Contract-Relax, Hold-Relax, and Slow Reversal-Hold-Relax. While a technique that achieves relaxation of the antagonist presumes that, secondarily, facilitation of the agonist will occur, the relaxation techniques are most useful for initial relaxation. They become substitutes for the "passive stretching" that has been used traditionally to increase the passive range of motion. Their use will not promote the development of strength and endurance of the agonist.

The specific techniques are rarely used singly; they are usually combined in a sequence which promotes the desired effect. The use of techniques which employ the antagonist, reversal and relaxation techniques, is, more often than not, followed by the use of repeated contractions in order to emphasize movement in the desired direction. An arbitrary guide to selection of techniques cannot be given. The diversity of patients' problems, the degree of involvement, and the presence of pain are factors which influence selection. The Summary of Techniques, Table 1, provides certain guidelines as to indications and contraindications.

The battery of specific techniques permits choice as to the way or ways in which the basic procedures may be supplemented and adapted to the patient's needs. The patient's needs will be fulfilled most readily by combining basic procedures and specific techniques all of which become more effective when superimposed upon the spiral and diagonal patterns of facilitation, as individual patterns, as combining patterns, and as total patterns of movement and posture.

Basic procedures

MANUAL CONTACTS

Treatment of the patient frequently involves manual contact with the patient by the physical therapist. Manual contacts use pressure as a facilitating mechanism (ref. 8). This may be demonstrated very simply in the normal subject. When elbow flexion is resisted with pressure over the biceps brachii, the subject will be able to pull with effectiveness and strength and will be able to hold adequately in the shortened range of the motion. If a comparable degree of pressure is applied to the triceps as the subject flexes the elbow against resistance, he will be felt to contract the biceps less effectively and to hold less well against resistance in the shortened range of elbow flexion. This test will be most convincing when the amount of resistance given allows the range of motion to occur slowly with maximum effort by the subject.

The specific manual contacts described for the patterns employ pressure to the skin overlying the groups of muscles, tendons, and joints responsible for the patterns and in line with the direction to which resistance is to be applied. Since the direction of the movement is diagonal, the therapist assumes a position in the diagonal. In this way, therapist and subject move together. If the therapist fails to move, or assumes a position away from the diagonal, the subject's efforts will be hindered and his movement may be distorted. The physical therapist takes hold of the subject and applies pressure purposefully but without producing painful stimuli. Painful stimuli may elicit a withdrawal response which does not facilitate voluntary, controlled motion in the desired direction.

Manual contacts may be varied somewhat with shifting pivots of emphasis and in special situations in order to make pressure more specific in relation to a given muscle or group of muscles, i.e., in the axilla as a stimulus to the subscapularis when internal rotation is to be emphasized in the lengthened range of extension–adduction–internal rotation of the upper extremity. Manual contacts may be used to place a demand or they may be used to give the patient security. The latter applies when treating patients who have symptoms of pain. The contacts should then include pressure for both antagonistic and agonistic patterns in order to provide security.

Pressure may be used as a sensory cue to help the patient to understand the direction of the anticipated movement. For example, if a patient is to flex his neck with rotation toward the left, briskly tapping him on the chest above the left nipple will guide his movement.

Suggestions for Learning Manual Contacts

Practice placing hands for a specific pattern. Have subject move actively through the available range of motion.
a. Did the positions of your hands allow the full range of motion to occur or did they impede the range of motion? Did you move as the subject moved? Was your body in line with the diagonal direction of the movement?
b. Were your hands accurately placed so that the pressure was over the muscle groups, tendons and joints participating in the movement?

COMMANDS AND COMMUNICATION

Communication with the patient relies upon sensory cues. Manually contacting the patient's skin, telling the patient what to do, and engaging vision to give movement direction are means of communication which demand the patient's attention. Skin can make both temporal and spatial discriminations, vision is the great spatial sense, while audition is the great temporal sense (ref. 4).

Verbal commands place a demand upon the subject. In order to provide an adequate stimulus, the developmental level of the subject and his ability to cooperate must be considered. Commands given to the lucid, normally innervated adult subject will obviously be very different from the motivation employed with a child of six months. Where the developmental or cooperative level is low, adequate stimulation may be more dependent upon physical demands and visual cues than upon verbal commands.

Tone of voice may influence considerably the quality of response (ref. 1). Strong, sharp commands simulate a stress situation and are used when maximal stimulation of active motion is desired. Overuse of strong commands may result in adaptation by the subject, and they should be reserved for demands for further effort. Moderate tones of voice should be used when the subject is responding with his best effort, and in giving preparatory commands. Soft tones of voice are desirable when security is needed by patients who have symptoms of pain.

Preparatory commands must be clear and concise. They may be made more meaningful by demonstration of the desired movement and by providing a visual cue. The therapist may point at an appropriate object on which the patient is to focus his attention. Using vision to lead movement encourages response. A child is frequently led to perform by enticing him to look at and pursue an object. In the adult vision also reinforces movement and should be utilized with preparatory commands. Complete understanding of what is to be done is most important when pain is a factor. When understanding and confidence have been gained, when patterns have been learned and pain is relieved, preparatory commands play a lesser role.

Action commands must be short, accurate, and timed to the physical demands. "Push" or "pull" are commands for isotonic contractions. "Hold" is the command for isometric contraction. "Relax" or "let go" are the commands for voluntary relaxation. The timing of action commands is extremely important. A premature command results in poor initiation by the patient and loss of control of the motion by the physical therapist. Delayed commands result in a lessened response especially where a stretch is being used. The best use of verbal stimulation requires the attention of the subject and the physical therapist. This precludes idle chatter while working, since few individuals are able to put forth maximum effort and converse simultaneously. During intervals of inactivity conversation is desirable since it provides a change in activity which has a recuperative value.

Self-evaluation may include the following questions:

1. Did my hands and my words help the subject to understand what was expected of him? Did I rely too much on too many words, rather than on careful use of my hands?
2. Did I command him to "push" or "pull' when I meant for him to hold?
3. Did he perform in accordance with normal timing?
4. Did my commands encourage normal timing?
5. Were my commands directed at the pivot of emphasis?
6. Would the patient's response have been better if I had used a stronger tone of voice? If I had asked him to look in the direction of the movement, would he have performed more adequately?

STRETCH

Stimulus

That muscle responds with greater force after stretch has been superimposed is a fact of physiology (ref. 10). Stretch, for this reason, may be used as a stimulus.

In order to achieve a stretch stimulus in any given pattern the part must be placed in the extreme lengthened range of that pattern, which is the completely shortened range of the directly antagonistic pattern. This is the starting position of the pattern as portrayed by the black figure of the physical therapist in the illustrations of individual patterns. All components of the pattern must be considered to obtain a stretch stimulus. The component of rotation receives first and last consideration since it is the rotatory component which elongates the muscle fibers of the muscles in a given pattern. The part should be taken to the point where tension is felt on all muscle components of a given pattern.

Stretch Reflex

Once the position has been achieved with the stretch stimulus, a stretch reflex can be superimposed on the pattern. The reflex can be elicited manually by "quickly" taking the part past the point of tension being certain that all components are stretched especially proper rotation. At the exact same moment the reflex is elicited the patient attempts the motion. To be certain that the stretch and the patient's efforts are synchronized the command should be, "Now, push," or "Now, pull." This warns the patient to be prepared to attempt the motion.

Even in completely paralyzed muscles there may

be a contraction over the reflex arc when a stretch reflex is elicited (ref. 24). This contraction is followed by relaxation of the muscles stretched. Repetition of the stretch reflex with patient's efforts timed accurately is essential. The stretch reflex can be used to initiate voluntary motion as well as increase strength and enhance a quicker response in weak motions. Use of stretch reflex aids the patient with intact innervation to learn and perform the patterns with greater ease. Pain should always be avoided in using the stretch reflex. It is contraindicated with patients who have the problem of pain or with patients whose skeletal, joint or soft tissue structures should not be subjected to sudden movement.

When either stretch stimulus or stretch reflex is used, commands for voluntary motion are always used. By so doing, any potential for voluntary control will be tapped more readily. A proper balance of stretch reflexes is necessary to postural control. The stretch reflex may be used repeatedly as in the technique of repeated contractions.

The use of the stretch reflex should be judicious especially when stimulating flexion responses. Flexion reflexes may become dominant creating an imbalance between flexion and extension.

Suggestions for Learning

1. Stretch stimulus
   a. Practice placement of a part in the lengthened range of a pattern to the point of tension.
   b. Consider rotation at the proximal joint first, then all other components from proximal to distal.
2. Stretch reflex
   a. Practice taking the part to the point where tension is felt.
   b. "Quickly" move the part into and past the point of tension of components.
   c. Synchronize commands so that the physical therapist and the patient are working together.
3. Was rotation considered first and last?
4. Were the patient and physical therapist completely together?
5. Was pain avoided?

TRACTION AND APPROXIMATION

Traction, separating the joint surfaces, and approximation, compressing the joint surfaces, are directed toward the joint receptors. Joint receptors are responsive to alterations in joint position; the effect of joint receptor discharge upon motoneuron responses probably depends upon the position of the joint and upon the type of joint movement (ref. 1). In treatment, the use of traction seems to promote movement while approximation promotes stability or maintenance of posture.

Traction and approximation are used to stimulate the proprioceptive centers supplying the joint structures themselves. Either separating joint surfaces (traction) or compressing the joint surfaces together (approximation) are helpful means of stimulation. Manual contacts make possible the use of traction or approximation. In general, traction is used where the motion is one of pulling, and approximation is used where the motion is one of pushing. This is in keeping with normal activities. For example, if one attempts to lift a weight, the joint surfaces are separated by the weight unless the muscles contract to perform the job of lifting. In pushing a heavy object, the joint surfaces are approximated because of contact with the object and contraction of the necessary muscle groups. When there is marked weakness, traction or approximation are maintained throughout the active range of motion. Traction and approximation may be contraindicated in patients having acute symptoms. However, in patients who have arthritis, traction often encourages range of motion.

Approximation may be used to stimulate postural reflexes. To encourage sitting balance a suddenly applied pressure in a downward direction should be exerted on the shoulders. This maneuver will be most effective when the spinal column is in a nearly extended position. As the pressure is applied, the patient is commanded to "Hold," at which time resistance is applied anteriorly at one shoulder and posteriorly on the other. The patient attempts to hold himself as rigidly as possible, thereby preventing rotation of his trunk. To encourage standing balance, approximation may be given at the pelvis. Again, the alignment of joint structures is important. The pelvis should be tilted by gripping the brim on either side while the extremities are in extension. Pressure is then applied suddenly downward through the pelvis and extremities with command, "Hold." Resisting rotation at the pelvis as the patient sustains his position encourages stability.

During mat activities, postural responses in the upper extremities may be enhanced by superimposing pressure downward on the scapulae as the patient maintains himself in a creeping posture.

Approximation is always applied in resistive walking except where weight-bearing is contraindicated.

Approximation is most effective if it is applied alternately as the patient puts weight on his lower extremities during the stance phases. This can be done by downward pressure applied through the shoulders or the pelvis.

Suggestions for Learning Use of Traction and Approximation

1. Practice applying traction and approximation using specific manual contacts.
a. Did the subject perform with greater ease and strength?
b. Were traction and approximation applied without producing pain?

MAXIMAL RESISTANCE

Movement performed against resistance of sufficient degree to demand maximal effort produces an increase in strength (ref. 10). When maximal effort is demanded, the amount of resistance used may be termed maximal. Performance with maximal effort may be hazardous to some patients (ref. 10). The hazard is perhaps greater when mechanical resistance is used than when manually applied resistance is appropriately graded to the patient's effort. The Valsalva phenomenon can best be avoided by allowing the patient to move when the command for movement is given rather than requiring a prolonged effort to overcome the opposition provided by the physical therapist. Where sustained effort by the patient warrants precaution, prolonged repetition of isometric contractions, holding, should be avoided or used guardedly.

Maximal resistance as applied in techniques of proprioceptive neuromuscular facilitation may be defined as the greatest amount of resistance which can be applied to an isotonic or active contraction allowing full range of motion to occur. When applied to an isometric contraction, maximal resistance is the greatest amount which can be applied without defeating or breaking the patient's hold. It therefore becomes necessary for the physical therapist to feel and sense the ability of the patient and to grade resistance accordingly. If these amounts of resistance were measurable, there would be a wide range of variation. The skill with which a physical therapist applies the manual contacts, superimposes pressure, stretch, traction, or approximation on the pattern will affect the amount of resistance a patient can overcome. Many mechanical factors involving the relationship of levers, the axis of motion, and

the effect of gravity play a role in determining the amount of resistance given but there is only one criterion for judging maximal resistance. This criterion is that the patient must put forth maximum effort but must be allowed to move the part slowly but smoothly throughout a range of motion. When isometric contraction is resisted, the patient must again put forth maximum effort but the physical therapist must not break or defeat the hold. Rather, the patient's ability to hold is developed by gradually increasing, from minimal to maximal, the amount of resistance in keeping with the patient's response.

Resistance may also be graded so as to encourage speed and repetition of a motion, conducive to the development of endurance. Grading resistance for this purpose demands skill and perception on the part of the physical therapist. Here the objective is to allow the patient to repeat the motion as many times as possible whereas in developing power throughout the range of motion, resistance is given so strongly that the patient may only succeed in a few repetitions of the pattern. As always, the needs of the patient determine the method. Patients who have normal innervation but with acute symptoms require very careful grading of resistance in specific, limited ranges of motion. The resistance used may be very slight but may be maximal for the patient's needs.

In treating patients who have deficiency of innervation, maximal resistance is one of the most important techniques when superimposed upon patterns of facilitation. It is maximal resistance which provides the means for securing overflow or irradiation from more adequate to less adequate patterns of movement of the head and neck, trunk, and the extremities. Stronger muscle groups within a pattern and stronger patterns must be utilized to augment the response of weaker muscle groups and of weaker patterns through a process of timing. The timing of application is coupled with appropriate gradation of maximal resistance.

Suggestions for Learning Maximal Resistance

Position the part in the lengthened range of a pattern with proper attention to all components of the pattern. Check manual contacts for accuracy, apply stretch, traction or approximation and command the subject to pull or push. Grade resistance so that the subject performs smoothly and slowly through the available range of the pattern.
a. Did the subject perform through the full range of the pattern?

b. Did rotation enter the motion first? Did the line of movement become diagonal following initiation of rotation? Was performance in the "groove" of the pattern?
c. Did the manual contacts allow the subject to move all component parts through their full range of of motion?
d. Did the distal parts move first?
e. Did pressure, stretch, traction, or approximation produce pain?
f. Were commands timed with manual demands and performance?

NORMAL TIMING

Normal timing is the sequence of muscle contraction which occurs in any motor activity, resulting in a coordinated movement. The importance of timing is realized in life when the normal subject attempts to learn a sport skill which involves a high degree of coordination. Routine activities of life involve timing and are learned through trial and error (ref. 9). Witness the baby learning to feed himself—he may open his mouth, lift the spoon, then close his mouth before emptying the spoon or he may empty the spoon before closing his mouth. The baby is in the process of learning the timing that establishes coordinated movements related to feeding himself.

In the normal process of development proximal control is evident before distal control. After coordinated, purposeful movements have been acquired, timing or sequential contraction of muscles occurs from distal to proximal. An example of this difference is that of the approach to rolling from supine to prone by the infant as compared with the coordinated individual. The infant makes his initial attempts by using neck and trunk motions and then learns to use the extremities effectively. The older child or adult will automatically use the extremities by placement and as reinforcement to assist in the process of rolling.

Distal to proximal timing is in keeping with the fact that the distal parts, the hands and feet, receive most of the stimuli for motor activities. Motions of the trunk that are proximal follow motions of the neck and extremities. For example, an article is grasped and lifted with action occurring first in the hand and proceeding to the elbow, shoulder, neck, trunk, and other extremities in accordance with demand.

Normal timing in patterns of facilitation may be demonstrated simply by superimposing resistance in the lengthened range of a specific pattern. The normal, coordinated subject will initiate the pattern by rotating the part first and then accomplishing the other components of the pattern from distal to proximal. Rotation always enters the motion first, and the diagonal direction of the pattern is controlled by the physical therapist through control of the rotation component. If rotation is not allowed to occur, the other components of motion cannot occur. After rotation is allowed to enter the motion, the normal subject will proceed to move first the distal pivots and action proceeds from there to the proximal pivots.

For example, if resistance is applied to the lengthened range of the flexion–adduction–external rotation pattern, the subject will grasp and rotate the entire extremity, the fingers flex and adduct toward the radial side, the wrist proceeds to flex toward the radial side, the forearm supinates, the shoulder externally rotates. After this initial rotation has taken place, the distal parts complete their range of motion by the time the elbow has flexed to the middle range; the shoulder proceeds to flex and adduct. The elbow completes its range of motion before the shoulder has completely flexed and adducted.

Normal timing may be prevented by excessive resistance to the rotation component and to the distal pivots. If the fingers and wrist are not allowed to move, action cannot occur at the proximal pivots.

If normal timing in patterns of facilitation has not not been developed or is deficient, it becomes one of the goals of treatment. Proximal deficiencies are corrected first in line with the normal process of development. Distal deficiences are corrected after proximal control has been established. If proximal control is adequate, distal control must receive emphasis in accordance with normal timing.

REINFORCEMENT

Techniques of proprioceptive neuromuscular facilitation employ reinforcement as a means of increasing the strength of a response. The major muscle components of a specific pattern augment and reinforce each other in order that the motion may be accomplished. Reinforcement extends beyond the specific pattern when maximal resistance is superimposed. An extremity pattern performed against resistance may demand reinforcement by the neck, trunk, and all other extremities. Combinations of patterns performed against resistance simulate stress situations which call into play the basic reflexes as reinforcement.

Timing for Emphasis

Timing for emphasis is based upon Beevor's axiom that the brain knows not of individual muscle action but knows only of motion. The urge of the subject to accomplish a motion brings into action the muscles necessary to the performance of that motion (ref. 5).

In timing for emphasis, maximal resistance is superimposed upon patterns of facilitation with due regard for normal timing in order that overflow or irradiation may occur from stronger to weaker major muscle components. The stronger muscle components must augment the weaker components; a weaker component cannot augment the response of a stronger component. Timing for emphasis provides the means for increasing response and stimulating action at a specific pivot within a pattern, a specific component in relation to that pivot, and a specific part of the range of motion of that pivot (ref. 18).

Timing may be accomplished by using either the stronger distal parts or the stronger proximal muscle groups. The process of timing produces irradiation from one group of muscles to the other. It may be done in the following ways:

1. By preventing the stronger component motions of a pattern from moving through any appreciable range with manual maximal resistance being applied; or,

2. By allowing the part to move against maximal resistance to the strongest point in the range of motion at which point a maximal isometric or "hold" contraction is done. After the "hold" is completed the patient is asked to "pull" or "push" strongly with no joint motion being allowed except in the weaker or joint needing emphasis.

If wrist extension toward the radial side is the weakest component of the pattern and needs emphasis, the physical therapist could promote irradiation by resisting strongly the pattern of flexion-abduction-external rotation of the shoulder allowing motion only to occur at the wrist joint. The resistance to wrist extension toward the radial side could be minimal or guided resistance to give the patient the feeling of motion occurring.

If the wrist extension toward the radial side is the strongest motion of the flexion-abduction-external rotation pattern, the physical therapist could apply enough resistance to prevent motion of the wrist joint from occurring but allow range of motion to occur in the shoulder joint. Resistance must be given slowly to allow maximal "build-up" if irradiation is to occur.

Timing may also be done in the two ways described by using the patterns of the stronger extremity to produce irradiation to the opposite extremity or extremities, such as arm or leg patterns or a combination of both. When these are used, maximal resistance is always given to the strongest part first. The weaker extremity is then guided and resisted into the desired pattern. Any strong pattern can reinforce another on the opposite side. They do not have to be corresponding motions on the two sides. Strong extensor patterns can promote irradiation to flexors on the opposite side as well as to the extensor groups.

Combining Patterns

Normal motor activity requires innumerable combinations of motions. Various segments of the body cooperate in an integrated fashion in order that movement may be coordinated and purposeful. This cooperation of various body segments may be termed reinforcement, since it is necessary to successful performance. In stress situations or in activities referred to as large muscle activities, such as active sports or manual labor, reinforcement becomes readily evident. Reinforcement in the mature normal subject occurs automatically and in accordance with the demands of the situation. Under stress, irradiation or spread of muscle activity occurs automatically in other body parts so as to support the desired movement or activity (ref. 10).

Reinforcing motions are acquired in the developmental process and in the learning of functional skills. They become established at the reflex level and include basic reflexes such as the tonic neck and labyrinthine reflexes, the primitive mass flexion and mass extension reflexes, and the postural and righting reflexes. These reflexes play a role in the automaticity of reinforcement in stress situations.

The relationships between vision and movement provide further keys to reinforcement. In normal subjects, vision is important to perception and performance of a motor act (ref. 25).* As head and neck and upper trunk patterns are performed, vision leads the movement. Asking the patient to look in the direction of the movement will contribute to his performance. As the upper extremity patterns are done, the eyes may follow the hand, or, again the patient may be asked to look in the desired direction so that the hand follows the eyes. Children

* The suggested readings include references to the works of Hellebrandt and coworkers, and others on involuntary patterning evoked by exercise stress, on simultaneous static and dynamic exercise, on cross education and indirect learning, and on the effects of combining movements during exercise.

are lured to look and pursue. Adults may require similar stimuli.

The normal subject is capable of performing all combinations of patterns of facilitation and the potentials for reinforcement are many. Reinforcement is a two-way proposition depending upon the demands of the situation.

The patterns of the neck may reinforce the trunk, of its opposite extremity. Flexion of a lower extremity reinforces adduction of an upper extremity. Extension of a lower extremity reinforces abduction of an upper extremity. The pattern combinations for reinforcement are given in Tables 3–9, pp. 209–212.

The deficient neuromuscular mechanism is incapable of meeting the physical demands of life.

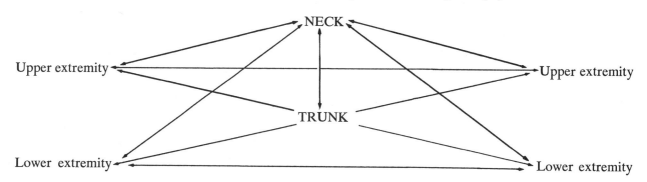

or the trunk may reinforce the neck. The neck and trunk may reinforce the unilateral or bilateral extremities, or the extremities may reinforce the neck and trunk. The unilateral upper extremity patterns which reinforce the neck, or are reinforced by the neck, are those in which the eye may readily follow the hand. The extremities may reinforce each other with bilateral symmetrical, bilateral asymmetrical, or bilateral reciprocal combinations. One extremity may reinforce its ipsilateral or contralateral upper or lower extremity. A sample of bilateral extremity patterns is charted:

While the potentials for reinforcement are

This does not preclude the reinforcement of weak patterns by stronger patterns. The less the deficiency, the more effective reinforcement will be. The greater the deficiency, the greater the need for reinforcement.

Selection of pattern combinations for reinforcement must consider the developmental level of the subject. The infant who has not reached the level of reciprocal motions cannot be expected to respond to reciprocal patterns. Bilateral symmetrical and asymmetrical patterns must first be established, along with mass flexion and mass extension of neck, trunk, and extremities.

Extremity and pattern to be reinforced	Patterns of opposite extremity used as reinforcement			
			Reciprocal	
	Symmetrical	Asymmetrical	Same diagonal	Opposite diagonal
Lower extremity				
Flexion— adduction— external rotation	Flexion— adduction— external rotation	Flexion— abduction— internal rotation	Extension— abduction— internal rotation	Extension— adduction— external rotation

many, certain combinations of the major components are used. Flexion, extension, or rotation patterns of the neck reinforce the homologous trunk patterns. Flexion of the upper extremities reinforces upper trunk extension. Extension of the upper extremities reinforces upper trunk flexion. Flexion of the lower extremities reinforces lower trunk flexion and extension of the lower extremities reinforces lower trunk extension. Flexion or extension of one extremity reinforces flexion or extension

The adult who presents gross deficiency of the neuromuscular mechanism must begin recovery with the same fundamental patterns that he used in infancy. Response in the child and the adult may be stimulated by resisted rolling, pivoting, creeping, and plantigrade walking. Activities such as coming to a sitting position, push-ups, getting to a kneeling position, getting to a standing position may be resisted in order to accelerate the learning process. Resistance to sitting blance, balancing in the hand-knee position, kneeling balance, and standing balance brings into play the postural and righting reflexes which augment the strength of response. Other techniques of facilitation including timing for emphasis, reversals, rhythmic stabilization, and repeated contractions may be superimposed.

Whereas the use of developmental activities increases the response of the neuromuscular mechanism, refinement of response and function is dependent upon specific reinforcements. A specific pattern which is deficient is reinforced by a related pattern. The relationship may be functional and topographical. Functional relationship exists between the bilateral patterns of the upper extremities when they are combined with upper trunk flexion or extension in motions such as chopping and lifting. Topographical contiguity exists, for example, between the flexion–abduction–external rotation pattern of the upper extremity and the neck extension with rotation pattern, when this pattern is performed toward the side of the extremity. The trapezius muscle provides their topographical contiguity.

The specific pattern of reinforcement must present greater total ability than the pattern it is to reinforce. For example, the extension–abduction–internal rotation pattern is a good choice of reinforcement for the flexion–adduction–external rotation pattern of the opposite extremity. However, if it does not enhance the response of the less adequate pattern, a second pattern must be selected.

Decision as to desirable patterns of reinforcement is partially based upon the correction or prevention of imbalances. Since the potentials for reinforcement are so numerous, it should be possible to prevent increase of imbalances. Wise selection of reinforcements may serve a dual purpose. A reinforcement may strengthen the response in a weaker pattern and may, at the same time, be a desired pivot of emphasis. For example, the flexion–adduction–external rotation pattern of the left upper extremity may be deficient. The extension–abduction–internal rotation pattern of the right upper extremity may provide good reinforcement even though it may be less strong than its antagonistic pattern. Using this reinforcement benefits the reinforcing pattern as well as the reinforced pattern.

The use of various related patterns as reinforcement for a deficient pattern is advisable, since in normal activity many combinations of motion are used. A specific pattern should be re-educated and trained to work in combination with its related patterns. Such a procedure helps to reestablish the automaticity of reinforcing motions.

The technique of reinforcement is comparatively simple once the basic techniques have been mastered. The part to be reinforced demands meticulous regard for manual contacts, stretch, and traction, and maximal resistance is of paramount importance. The pattern used as a reinforcement being stronger may be given slightly less regard. Where great weakness exists, the more skillfully both parts are handled, the more effective reinforcement will be. Both parts may be placed in the lengthened range of the desired patterns. The patient is instructed and commanded to "push" or "pull" while the therapist controls the situation by allowing the stronger pattern to respond first, but resisting it strongly enough to encourage the weaker pattern to respond. The patient's urge is to accomplish the motions, and by resisting the stronger motion, the weaker motion derives stimulation.

EXAMPLE OF TECHNIQUE

If it is desired, for instance, to reinforce the flexion–adduction–external rotation pattern of the right lower extremity with the extension–abduction–internal rotation pattern of the left lower extremity, with emphasis on the right hip, the lower extremities are placed on stretch for these respective patterns. The hip is the pivot of emphasis, and since reinforcement is necessary because of weakness, the manual contacts and position of the physical therapist may be varied to allow for better control of the two extremities. The physical therapist may grip the patient's heels, since rotation may be easily controlled from this point. The left lower extremity is allowed to move first, and as the strength of the response is felt, the right lower extremity is allowed to move. When the right lower extremity has moved through as much range as possible, and the reinforcing extremity has moved to its strongest part of the range, the patient is asked to "hold." The therapist then "equals" the "hold" on both extremities, that is, he resists all components of both patterns as strongly as possible, without derotating and defeating the "hold." At this point, the physical therapist may shift to any other selected technique, such as repeated contractions, slow reversal, slow reversal–hold, or rhythmic stabilization, in order to further facilitate the desired motion. The element of timing exists in the reinforcement of one pattern by another pattern, just as it does in relation to reinforcement of weaker components of a single pattern by stronger components of the same pattern.

Partial reinforcement may be achieved by having the patient perform a related pattern actively and simultaneously, such as motions of the neck, gripping the table when a unilateral upper extremity motion is being emphasized, or gripping the table with both hands when lower extremity or lower trunk motions are performed. When upper trunk motions are performed, the subject must not grip the table—this will disturb the upper trunk patterns, which are closely related to the upper extremity patterns. Moving the upper extremities actively through related patterns while upper trunk motions are performed may provide some reinforcement for the trunk motions.

Recuperative Motion

Recuperative motion is the use of a new combination of movements to reduce or circumvent fatigue produced by repetitive activity against resistance (ref. 12).

Fatigue resulting from repeated or strenuous physical activity is a well recognized and accepted phenomenon in normal life. The normal subject recognizes and circumvents the fatigue factor in his performance of physical tasks requiring effort. He knows that if he wants to do something, he does not tire as readily as he does when performing a disliked task. He knows through experience that he can improve his endurance for a certain activity by working to the point of fatigue. He knows that while he may tire of a certain task or tire physically in performing it, he may perform for a longer period by changing the routine procedure of the task itself or by changing to another activity for a brief period of time.

A simple example is that of the normal subject who sets himself the task of waxing a piece of furniture. He wants to do the job, or someone else wants him to do it, to the extent that he is provided with a motive. Having started with determination he may find that very soon his right arm is tired. If he shifts from short strokes to longer strokes, it seems to help in relieving the fatigue momentarily. If he emphasizes pulling motions rather than pushing motions he may lengthen his period of performance. If he changes the position of his body or reinforces his arm motions with his trunk or other extremities, he may again lengthen his work period. Or he may shift from the use of his right hand to the use of his left hand and find some relief of fatigue. Finally, he may change to a completely different task, and later, returning to the

waxing job, find he is again able to perform adequately.

The task described requires many combinations of motions involving many parts of the body. When a certain combination of motions becomes fatiguing, the normal subject shifts the emphasis to another combination which may accomplish the task. He reinforces one combination of motions with other combinations and proceeds to vary his own activity in order to work for a longer period of time. This shifting of emphasis to other combinations of motions amounts to recuperative motion in the normal subject.

Techniques of proprioceptive neuromuscular facilitation employ recuperative motion in order to reverse fatigue factors. The reversal of antagonist techniques may be used as recuperative motion and the use of several different patterns of reinforcement constitute recuperative motion. When recuperative motion is employed, the patient will be able to work at a specific pattern which is in need of emphasis, for a longer period of time, thereby improving the strength and endurance in relation to that specific pattern.

A patient having weakness of elbow extension in the extension–abduction–internal rotation pattern of the upper extremity may fatigue readily in the performance of repeated contractions emphasizing elbow extension. Reversing the pattern to flexion–adduction–external rotation with elbow flexion may make it possible for the patient to repeat elbow extension in the initial pattern. If the elbow extension was being performed unilaterally without reinforcement of another stronger pattern, resisting the pattern bilaterally symmetrically, bilaterally asymmetrically, reciprocally in the same diagonal or reciprocally in the opposite diagonal, or resisting in combination with neck rotation to the same side may enhance the ability to continue repetitions of elbow extension. Further recuperation may be brought about by shifting to another pattern which requires emphasis in the trunk or other extremities, making it possible to again return to the initial pattern emphasizing elbow extension.

By following such a procedure, in view of the many combinations of motion available for reinforcement, it becomes possible to place an intensive demand on a specific, desired combination of motions. It is by such a procedure that the fatigue factor is circumvented in the application of techniques of proprioceptive neuromuscular facilitation.

Specific techniques—Voluntary effort

REPEATED CONTRACTIONS

Repeated excitation of a pathway in the central nervous system promotes ease of transmission of impulses through that pathway (ref. 18). Repetition of activity is necessary to the learning process and to the development of strength and endurance. Repeated activity of the weaker components of a pattern is obtained through a technique of emphasis, repeated contractions. In order to emphasize the response of a weak component of a pattern, or a weak pattern, motion is repeated until fatigue is evident in the performance of that motion. Fatigue will be delayed, and response will be enhanced if the stretch reflex is coupled with the patient's voluntary effort to initiate movement.

The less advanced form of repeated contractions involves only isotonic contractions stimulated by use of the stretch reflex as the patient attempts the movement. The response to stretch must be resisted by the therapist so as to enhance voluntary response and motor learning. Repeated isotonic contractions induced by stretch reflex may be the only choice when a patient cannot move voluntarily, or if the patient is unable to perform a "hold" with isometric contraction. When the stretch reflex is used, care must be taken to avoid creation of imbalances between flexor and extensor reflexes. Therefore, the skill lies in the ability to resist the response from stretching the muscle groups in the pattern and to time the use of resistance with the patient's voluntary effort.

The verbal commands are combined with stretch. That is, as stretch is given, "Now," is synchronized with the maneuver, and "Pull!" follows immediately if flexion is being stimulated. For extensor movements the command becomes, "Now, (stretch) push!" "Now, (stretch) push!"

The more advanced form of repeated contractions utilizes both isotonic and isometric contractions. If a patient is only able to perform the simpler form, the advanced form becomes a goal of treatment. That is, the therapist must keep in mind the need to teach the patient to "hold" and must not be satisfied by use of only isotonic contractions. As the patient's strength and endurance improves, performance of a "hold" contraction may become possible. The more advanced form is executed as follows.

After the subject has performed initially against resistance with resultant overflow to a weaker pivot of action, he is instructed to "hold," with an isometric contraction, at the point where active motion is felt to be lessening in power. The physical therapist then secures the "hold" by resisting all components of the pattern in turn from distal to proximal. Resistance is maximal but the goal is to encourage the subject to "hold" rather than to defeat or break the "hold." When the entire part has become secure, or as it were, a complete unit, by the patient having sustained his effort, the physical therapist maintains resistance to all pivots of action equally and then resists more strongly at the weaker pivot of action. At the moment that resistance is increased at the weaker pivot, the physical therapist instructs the subject to "pull again" or "push again," thereby shifting from an isometric to an isotonic contraction. Having asked the subject to "push" or "pull" it is again necessary to grade the resistance so that active motion of the weaker pivot may occur.

By repeating isotonic contractions, the patient may be allowed to work toward the shortened range of the pattern. If the initial response was secured in the shortened range, the response may be channeled toward the lengthened range by gradually decreasing the range of motion through resistance.

Example of Technique-Isotonic and Isometric

PATTERN TO BE EMPHASIZED

Flexion–adduction external rotation of right upper extremity (Fig. 13).

PIVOT OF EMPHASIS

Action to be emphasized; elbow flexion.

Lengthened range of pattern (stretch range).

TIMING FOR EMPHASIS OF ELBOW FLEXION (BICEPS BRACHII)

Resist strongly the fingers and wrist flexion to the radial side, the supination of the forearm, and the external rotation of the shoulder. Allow beginning rotation to occur at the fingers, wrist, forearm, and shoulder, but do not allow full range of finger and wrist flexion toward the radial side and shoulder flexion and adduction to occur until the elbow begins to flex. As the elbow begins to flex, allow the distal components to complete their range of motion (normal timing) and then allow beginning range to occur at the shoulder.

COMMANDS TO THE SUBJECT

1. For initiation of active motion with isotonic contraction: "Squeeze my hand, turn it and pull it up and across your face. Bend your elbow!" Patient pulls while the physical therapist resists the motion as outlined above.

2. For securing isometric contraction in preparation for repeated isotomic contractions: "Hold it!" "Hold" command is given when patient has achieved maximum elbow flexion. All major muscle components are resisted without defeating the hold at any pivot of action. The part becomes secure. Derotation is avoided.

3. For repetitions of elbow flexion: "Now, pull," and "pull," "and pull," "and pull again!" "And rest." The physical therapist having felt the subject "hold" throughout the pattern, resists slightly the sustained "hold" contraction at the elbow, proceeds to give the commands for repetitions of elbow flexion. Repeated application of resistance immediately precedes the commands. Increased range of motion is allowed to occur.

Repeated contractions technique is indicated where weakness and incoordination are primary problems. It is contraindicated where sustained effort is contraindicated and in very acute situations.

Correction of Imbalances

Repeated contractions with timing for emphasis is the technique of choice in the correction of imbalances. In patterns of facilitation imbalances may present themselves with relation to the major muscle components of a specific pivot. For example, a weak anterior deltoid may be overpowered by a strong clavicular portion of the pectoralis major. Both muscles are related in the flexion–adduction–external rotation pattern. When this pattern is performed, response in the anterior deltoid is dependent upon maximal resistance to the stronger clavicular portion of the pectoralis major and to the component of external rotation. Emphasizing external rotation will prevent excessive adduction of the shoulder and will stimulate the anterior deltoid.

Imbalances may also occur in relation to the major muscle components of related patterns with reference to a specific pivot. For example, the ulnar flexor of the wrist may be stronger than the radial flexor. When the flexion–adduction–external rotation pattern is performed for shoulder emphasis, the wrist may fail to proceed to radial flexion and this will distort the pattern with regard to the proximal components. For emphasis of the shoulder pivot in a situation of this kind, the wrist must be guided into radial flexion and supination. For correction of the imbalance at the wrist, radial flexion must be considered a pivot of emphasis, and external rotation of the shoulder and supination of the forearm must be resisted strongly so as to stimulate the radial flexor. Correction of proximal imbalances will enhance the correction of distal imbalances.

Imbalances may also exist between antagonistic patterns in relation to various pivots of action. For example, a patient who has strong rhomboids and teres major; and a weak anterior deltoid and clavicular portion of the pectoralis major has an imbalance in favor of extension–abduction–internal rotation of the shoulder. The same patient may have a strong biceps and a weak triceps resulting in imbalance in favor of the flexion–adduction–external rotation pattern with reference to the elbow. If he has a weak ulnar extensor of the wrist and a strong radial flexor, he again presents an imbalance in favor of the flexion–adduction–external rotation pattern with reference to the wrist. Such a picture presents imbalances not only between antagonistic patterns but within the components of both patterns. When techniques of facilitation are used, the stronger posterior scapular and shoulder muscles of the extension–abduction–internal rotation pattern must be used to stimulate the weaker triceps and ulnar extensor of the wrist. The distal power in the radial flexor of the wrist and the stronger biceps must be used to stimulate the weaker anterior deltoid and clavicular portion of the pectoralis major. Correction of imbalance at the shoulder deserves the first emphasis, but the imbalance at the elbow and wrist must also receive due emphasis.

Hyperactive Reflexes

Just as imbalances may exist between antagonistic muscle groups so may imbalances exist between antagonistic reflexes. For example, if extensor spasticity is so severe as to prevent a patient from voluntarily flexing his left lower extremity, repeated use of stretch reflex coupled with his voluntary effort may be helpful. Positioning the patient on his right side or in the hand-knee position will promote response because of the favorability of these positions for flexor response. In this way the patient may be able to produce more voluntary movement on which resistance can be superimposed. Increased voluntary response will help to reduce hyperactivity of the antagonistic reflex.

Pivots for Emphasis

Determination of imbalances and selection of pivots of emphasis are based upon careful evaluation of the patient's strengths and weaknesses. In techniques of facilitation, the development of power and the correction of imbalances proceeds from proximal to distal in line with the normal process of development. Proximal power and control is essential to stability and to the process of overflow or irradiation. Therefore, weakness and imbalance of neck and trunk musculature receive first emphasis. The proximal pivots of the extremities, the shoulder and hip, receive second emphasis, and the more distal pivots receive third emphasis. Where generalized weakness exists, proximal emphasis is necessary and precludes emphasis of a weak distal pivot. Weak distal pivots receive a certain amount of stimulation during proximal emphasis because all components of the pattern are considered. In techniques of facilitation it becomes futile to attempt to strengthen a nonfunctional anterior tibial, if the muscle groups related by pattern are equally deficient.

Rhythmic Initiation

Rhythmic initiation, or rhythm technique, is used to improve the ability to initiate movement. This technique involves voluntary relaxation, passive movement, and repeated isotonic contractions of the major muscle components of the agonistic pattern. Using rhythmic initiation is helpful to those patients who lack the ability to initiate movement because of rigidity (Parkinsonian) or severe spasticity. The patient is helped, too, to become aware of the direction of the movement. Those who are lethargic and are slow in their movement, the

elderly, and those who have diminished position sense may be stimulated and guided by this method. The technique does not involve forced movement. If only a limited range of movement is possible, this is the starting point. Movement should not produce pain; pain will only limit movement.

EXAMPLE OF TECHNIQUE

The physical therapist asks the patient to relax and "Let me move you." The therapist then moves the part through the available range of the pattern, giving attention to all components of the pattern with special attention to the distal parts. During the movement of the part, emphasis is in the direction of the agonistic pattern although the part is returned, of course, toward the shortened range of the antagonistic pattern. As relaxation is felt to occur and movement is more easily accomplished, the patient is commanded, "Now, help me just a little." Again, the emphasis is in the direction of the agonistic pattern. After the patient has assisted for several repetitions the physical therapist gradually superimposes resistance and increases the amount as the patient's response is felt to increase. He is commanded to "Pull!" or "Push!" as the occasion requires. After several repetitions against resistance, the patient is allowed to move actively by himself to sense the increased ease of movement.

REVERSAL OF ANTAGONISTS

Reversal of antagonists techniques are related to normal responses and good performance is indicative of normalcy of function.

In normal physical activity the reversal of antagonists plays an important role. The examples of sawing wood, chopping wood, rowing a boat, walking, running, grasping and releasing objects are trite but true illustrations of this phenomenon in life's activities. When antagonists fail to reverse in accordance with the demand of the activity, function is immediately impaired in relation to power, skill, or coordination. The objective of neuromuscular education or re-education and of therapeutic exercise may be said to be the development or redevelopment of a normal reversal of antagonists through a normal range of motion. This implies correction of imbalances and development of strength, coordination, and endurance.

The techniques are based upon Sherrington's principle of successive induction (ref. 18). The patient who does not respond well to reversal techniques and in whom the desired response can be achieved only by resistance in pattern with repeated contractions has severe involvement, i.e., the hemiplegic patient who presents a picture of disturbed patterns and who responds to reversal with an increase in spasticity also demonstrates a low level of functional participation. He is the patient who when his antagonists should reverse in a functional movement "bumps" into his spasticity, and his function is impaired thereby. For such a patient, patterns must be re-established with first emphasis on the proximal pivots, and the technique of repeated contractions is indicated. The reversal of antagonists becomes a goal of treatment rather than a technique of treatment.

The stimulation of the agonist by resisting a contraction, either isotonic or isometric, of the antagonist is readily demonstrated in the normal subject by determining the amount of resistance which can be overcome in performing a motion such as elbow flexion. If the motion of elbow extension is then performed against maximal resistance, a succeeding motion of elbow flexion should be performed more strongly, overcoming greater resistance than was possible on the initial attempt.

Techniques utilizing the reversal of antagonists are superimposed upon patterns of facilitation and with proper attention given to the manual contacts, maximal resistance, and timing of the pattern. The potentials of reversal techniques are several because of the variations possible. Either isotonic or isometric contractions may be used, or a combination of both types of muscle contraction may be used.

There are three reversal techniques which may be used primarily for the purpose of stimulation, as evidenced by a build-up in power or a gain in range of motion. Slow reversal involves an isotonic contraction of the antagonist, followed by an isotomic contraction of the agonist. Slow reversal–hold involves an isotonic contraction, followed by an isometric contraction of the antagonist, followed by the same sequence of contractions of the agonist. Rhythmic stabilization involves an isometric contraction of the antagonist, followed by an isometric contraction of the agonist and results in the co-contraction of antagonists. Detailed descriptions of the performance of the slow reversal technique follow.

Slow Reversal; Slow Reversal—Hold

A patient may present weakness of the flexion–adduction–external rotation pattern of the right lower extremity with reference to the hip, with good power in the antagonistic pattern of extension–abduction–internal rotation. If the antagonistic pattern is to be used as a a means of stimulation by using reversal, the physical therapist must be prepared to resist the antagonistic pattern, but will vary the manual contacts so as to be able to apply maximum proprioceptive stimulation to the weaker pattern. The procedure should be to require the patient to perform the agonistic motion of flexion–adduction–external rotation, the physical therapist applying optimum manual contacts with maximal resistance, in order to determine the response of the patient. The physical therapist then shifts his manual contacts to those described for the extension–abduction–internal rotation pattern and asks the patient to perform this motion against maximal resistance. The therapist then shifts again to the agonistic pattern of flexion–adduction–external rotation and determines if the patient performs with greater power or with increased range of motion. The grading of resistance is essential in order to demand a strong contraction of the antagonist and to allow range of motion to occur when the agonistic pattern is performed. The manual contacts must be varied to the extent that it is possible for the physical therapist to shift contacts and resistance smoothly, making it possible for the patient to shift smoothly and effectively from one pattern to the other. If the patient gains little range by reversing at the point where the agonistic pattern of flexion–adduction–external rotation had become ineffective, a reversal should be attempted from the lengthened range of extension–abduction–internal rotation, allowing the patient to work to the strongest part of the range of this pattern before requiring a succeeding contraction of the agonistic pattern. The reversal process may be repeated several times with the agonistic pattern being resisted last. When the agonistic pattern has been stimulated, the patient is instructed to "hold" immediately following the gain in flexion–adduction–external rotation, and repeated contractions may then be used as a technique of emphasis in order to gain further range and develop power or endurance in the agonistic pattern.

The series of commands for slow reversal, using extension–abduction–internal rotation to stimulate flexion–adduction–external rotation, could be as follows:

"Push your foot down and out toward me"—extension–abduction–internal rotation–isotonic–antagonist.

"And pull your foot up and across your body"—

flexion–adduction–external rotation–isotonic–agonist.

"And push down and out"—extension–abduction–internal rotation–isotonic–antagonist.

"And pull up and over"—flexion–adduction–external rotation–isotonic–agonist.

"And hold"—flexion–adduction–external rotation–isometric–agonist.

"And pull, and pull, and pull again"—flexion–adduction–external rotation–isotonic–repeated contractions for emphasis–agonist.

Slow reversal–hold requiring an isotonic and then an isometric contraction may be performed in the same manner, with a "hold" command inserted after each active command.

If slow reversal techniques are performed against maximal resistance an increase in range or build-up in power should be felt on each successive isotonic and isometric contraction. The physical therapist should remember that on a "hold" command the rotation is resisted strongly, but the hold is not defeated or broken.

Rhythmic Stabilization

Whereas slow reversal technique employs isotonic contractions, and slow reversal–hold employs isotonic and isometric contractions, a third technique of stimulation, based upon reversal of antagonists, is rhythmic stabilization. Rhythmic stabilization employs isometric contraction of antagonistic and agonistic patterns, which results in co-contraction of antagonists if the isometric contraction is not broken by the physical therapist. Because only isometric contractions are employed, rhythmic stabilization has a by-product of increasing circulation. Rhythmic stabilization performed on the normal subject results in a build-up of holding power so great that the hold cannot be broken unless rotation of the part is defeated. If the normal subject is instructed to hold the arm completely still or to hold it stiffly, he will do so by performing isometric contractions of all groups about a given joint, and the co-contraction of antagonists may be felt. If the normal subject thinks of holding his arm "up" or holding it "down," he will be felt to contract first one group, then the opposite group, with relaxation occurring between the isometric contraction. This is not rhythmic stabilization, for as the subject performs, he will be felt to shift into alternating isotonic contractions rather than to develop a co-contraction of antagonists. Careful grading of resistance to the "hold" contractions, with special consideration given to the rotation components makes it possible

for the subject to stabilize. The patient must not be defeated with so much resistance that he finds it necessary to contract isotonically in order to recover or maintain his position. This becomes especially important in treating patients who have normal innervation but problems of pain. The grading of resistance will be as accurate as the physical therapist's ability to feel the patient's response. In order to develop the patient's ability to stabilize, the physical therapist may move the part through a slight range of motion so as to better bring in the antagonistic patterns, but this must also be done by sensing the patient's response and without defeating the patient's hold in the rotation components.

Where there is inability to perform isometric contractions, as in ataxia, rhythmic stabilization may be impossible for the patient to perform. The patient must be taught to "hold." One approach is to have him perform slow–reversal–hold through decreasing ranges of motion until no motion occurs. If ataxia is severe, use of slow reversal with adjustment of decrements of range at various points, and attempts to hold at these points will be necessary.

EXAMPLE OF TECHNIQUE

Rhythmic stabilization may be performed at any desired point in the range of motion. When used to stimulate a weaker pattern in an attempt to gain further range of motion in that pattern, the procedure is as follows: Given weakness of the flexion adduction–external rotation pattern of the hip, the patient is required to pull through as much range in that pattern as possible. He is then instructed to hold the part still, and the physical therapist equals all components of the agonistic pattern with maximal resistance, but without breaking the hold. At this point the physical therapist shifts the resistance to all components of the antagonistic pattern without breaking the hold. Several repetitions of this procedure are then followed by a hold contraction of the agonistic pattern, and repeated isotonic contractions are used, allowing further gains in range of motion. The commands used could be as follows:

"Pull your foot up and across your body"—flexion–adduction–external rotation–isotonic–agonist.

"And hold it"—flexion–adduction–external rotation–isometric–agonist.

"Now hold it still; don't let me move it"—extension–abduction–internal rotation–isometric–antagonist.

"And hold"—flexion–adduction–external rotation–isometric–agonist.

"Now, pull, and pull, and pull"—flexion–adduction–external rotation–isotonic–repeated contractions–agonist.

Success in application of reversal of antagonists techniques will be dependent upon maximal resistance to the antagonistic pattern, control of antagonistic and agonistic patterns so that the motion may be performed smoothly, and careful grading of resistance between the stronger antagonist and the weaker agonist. Reversals may be performed in any desired part of the range of motion which gives the desired response. They may be performed through a full range of motion or a minimal part of the range of the antagonistic pattern, if that is adequate for stimulation of the agonist.

RELAXATION

A technique which demands contraction of one pattern of facilitation demands a lengthening reaction, relaxation, or inhibition of the directly antagonistic pattern. Any technique which demands or makes possible a gain in range of motion in one pattern has achieved relaxation of its antagonistic pattern. This relaxation, or inhibition, of the antagonist during facilitation of the agonist relies upon reciprocal innervation which was demonstrated by Sherrington (ref. 18). In the normal subject, the techniques of repeated contractions, slow reversal, slow reversal–hold and rhythmic stabilization, superimposed upon a pattern of motion, will stimulate the agonist and relax the antagonist. This may be simply demonstrated in normal subjects who have adaptive or postural shortening of the biceps femoris. If the subject is required to perform the flexion–adduction–external rotation pattern with emphasis at the hip and knee against maximal resistance, he may perform through greater range than when he does unresisted motion. If he is then required to perform repeated contractions, he may gain further range; if rhythmic stabilization is performed at the end-point of the range, he may again achieve added range. If at this point he is allowed to reverse through the antagonistic pattern of extension–abduction–internal rotation with maximal resistance, his performance in the agonistic pattern may be still further enhanced both in power and range of motion. Thus, stimulation and relaxation are inseparable.

The specific relaxation techniques are substitutes for passive stretching. The patient who seemingly has very little strength available may be able to produce a contraction of a shortened muscle of sufficient strength to encourage relaxation providing the contraction is skillfully resisted. Use of these techniques avoids painful reactions to stretch and are far less hazardous.

While relaxation techniques afford means to relaxation, the use of positioning to influence tonus, and maximum stimulation of related patterns may achieve more relaxation than will the application of a relaxation technique directed to a specific muscle group. For example, resisting the patient's efforts in creeping may bring about more relaxation of the extensors of the lower trunk than will contract–relax done in the supine position. These techniques of relaxation utilize maximal contractions of the antagonist followed by voluntary relaxation which, whenever possible, is followed by resisted contraction of the agonist.

Contract–Relax

In patients presenting marked limitation of range of motion with no active motion available in the agonistic pattern, some relaxation of the antagonistic pattern may be achieved by using contract–relax. This technique involves an isotonic contraction of the antagonist, allowing range of motion in rotation against maximal resistance but no range of the other components, followed by a period of relaxation.

EXAMPLE OF TECHNIQUE

The procedure would be to move the part passively into the agonistic pattern to the point where limitation is felt, and at this point, the patient is instructed to contract isotonically in the antagonistic pattern. The physical therapist resists the rotation as strongly as possible and then instructs the patient to "relax." It is necessary to lighten the pressure and to wait for relaxation to occur. Having felt the patient "let go," the physical therapist again moves the part passively, through as much range as possible, to the point where limitation is again felt to occur. The entire procedure is repeated several times, following which an attempt should be made to have the patient perform the agonistic pattern actively from the lengthened range. If the patient is not able to perform or initiate from the lengthened range, he may be asked to move actively in the direction of the agonistic pattern after each contract–relax procedure. However, as always, the goal is to initiate from the lengthened range and to perform throughout the range.

The commands for contract–relax in such a situation would be as follows:

"Pull your foot down and in"—extension–adduction–external rotation–isotonic–range of external rotation against maximal resistance–antagonist.

"Let go"—lighten pressure, support part, and wait for relaxation to occur, move the extremity into flexion–abduction–internal rotation–agonist.

Repeat procedure—then place the part in the lengthened range for flexion–abduction–internal rotation–agonist.

"Now pull up and out toward me"—flexion–abduction–internal rotation–isotonic–agonist.

"And hold"—preparation for repeated isotonic contractions.

"And pull, and pull, and pull, etc."—repeated isotonic contractions of agonistic pattern.

Hold–Relax

Hold–relax is a relaxation technique based upon maximal resistance of an isometric contraction. The technique is performed in the same type of sequence as contract–relax. Since an isometric contraction is involved, the command must be "hold" instead of "push" or "pull." Also, since no joint motion is implied, this technique may be performed as a method of achieving relaxation where muscle spasm is accompanied by pain. The isometric contraction must not be broken or defeated. In any acute situation, the technique should be demonstrated to the patient on a pain-free part. Exercise of the pain-free part has secondary benefits of general relaxation with reduction of pain, and, if resistance is maximal, irradiation to the painful area may occur without pain.

EXAMPLE OF TECHNIQUE

The fracture patient who has just had a cast removed may be helped to relax the part or to gain range with this simple technique. For example, a patient who has a united fracture of the head of the radius and is to perform active motion to encourage elbow extension has already established a mechanism of inhibition for elbow extension. By performing hold–relax to the biceps, with slowly increasing amounts of resistance applied to the isometric contraction, relaxation of the biceps may be achieved with resultant stimulation of the triceps. The part, of course, should be supported by the physical therapist, and after hold–relax is performed, the patient is instructed to extend the elbow without resistance. The commands in such a situation would be as follows:

"Just *hold* your elbow bent and don't let me move it"—resistance gently and slowly applied to supination using the distal contact of the flexion–adduc-

tion–external rotation pattern. Resistance is greater at the radially flexed wrist than it is at the elbow.

"Let go"—maintain gentle support of the extremity and wait for relaxation of the biceps to occur.

"Open your hand and push it down and away"—extension–abduction–internal rotation with elbow extension–isotonic without resistance.

Success of the technique will depend upon gently increasing resistance, encouraging the isometric contraction without defeating it, supporting the part while relaxation occurs, and having the patient move the part actively in the desired motion. The procedure may be repeated and followed with repeated unresisted active contractions of the agonist. Unresisted reversing movements, emphasizing the rotation, may also be used as a follow-up procedure.

Slow Reversal–Hold–Relax

Slow reversal–hold–relax is a technique involving an isotonic contraction of the range limiting pattern (the antagonistic pattern) followed by an isometric contraction of the antagonistic pattern, followed by a brief period of voluntary relaxation, followed by an isotonic contraction of the agonistic pattern. Relaxation must be achieved first at the exact point in the range where limitation presents itself. Maximal relaxation is dependent upon maximal resistance applied to the rotation component without allowing range of motion to occur in the other components of the antagonistic pattern.

EXAMPLE OF TECHNIQUE

If a patient presents limitation of active motion at $15°$ of flexion–abduction–internal rotation of the lower extremity with reference to the hip, that is the part of the range where relaxation of the extension–adduction–external rotation pattern must begin in order to stimulate the agonist and develop inhibition of the antagonist. The point at which the relaxation technique is to be performed is best determined by having the patient perform actively as much range of the agonistic pattern as possible, and then proceed with slow reversal–hold–relax of the antagonistic pattern. Using the manual contacts which are optimal for the antagonistic pattern, the patient is resisted so strongly as he attempts an isotonic contraction of the extension–adduction–external rotation pattern that no motion occurs, except in the component of rotation. The physical therapist instructs the patient to "hold," and resists the "hold" contraction with all the resistance applied to the rotation component. Having resisted the isometric contraction, the physical therapist instructs the patient to "relax," and at once releases his pressure, just supporting the part without moving it. As soon as relaxation is felt to occur, the physical therapist demands an isotonic contraction of the agonistic pattern, applying manual contacts for that pattern, but allowing the patient to flex, abduct, and internally rotate through as much range as possible. At this point, the slow reversal–hold–relax technique may be repeated, or if a definite gain in range has been achieved, the physical therapist may emphasize the recently gained portion of the range of motion by performing repeated contractions. Success in application of the technique will be dependent upon maximal resistance to the antagonistic pattern allowing only range of rotation to occur, decrease in pressure by manual contacts when the patient is instructed to "relax," and active motion against resistance following relaxation.

The commands for the technique as set forth in this example could be as follows:

"Pull your foot up and out toward me as far as you can"—flexion–abduction–internal rotation–isotonic–agonist.

"Now pull your foot down and in" and "hold" —extension–adduction–external rotation–isotonic–range of external rotation allowed and resisted as strongly as possible–isometric–antagonist.

"Relax"—lighten contact and shift manual contacts for optimal stimulation of agonist–flexion–abduction–internal rotation.

"Now pull up and out"—flexion–abduction–internal rotation–isotonic–agonist.

"And hold"—preparation for repeated contractions–isometric–agonist.

"And pull, and pull, and pull"—agonist–repeated isotonic contractions.

Reminders for learning

1. Learn the patterns as free active motion in accordance with normal timing. Begin with head and neck and upper trunk patterns with chopping and lifting. Practice with eyes leading the movement, and with eyes following the hands as the chop leads trunk flexion and the lift leads trunk extension. Proceed to upper extremity and then to lower extremity patterns. Perform the patterns in as many positions as possible including positions and postures of the developmental sequence. Analyze total patterns of movement and functional activities and identify their component patterns.
2. Learn to apply manual contacts accurately.
3. Practice giving commands with a normal subject performing active range of motion in accordance with normal timing.
4. Learn techniques of facilitation in the following order:
   a. Maximal resistance through full range of pattern in accordance with normal timing (Isotonic).
   b. Maximal resistance to "holding" in shortened range of pattern and various points in range (Isometric).
   c. Repeated contractions (emphasis of proximal pivot). Observe build-up in power or gain in range.
   d. Timing for emphasis (proximal, intermediate and distal pivots). Follow with repeated contractions.
   e. Slow reversal, slow reversal–hold, rhythmic stabilization. Work in various parts of the range of agonistic and antagonistic patterns. Observe build-up in power or gains in range of motion.
   f. Slow reversal–hold–relax, contract–relax, hold–relax. Observe relaxation and gains in active range and passive range.
   g. Reinforcement of one pattern by a related pattern. Practice performance of related patterns for various parts of the range and various pivots of action. See Table 3–9, pp. 209–212.
5. Practice all techniques with normal subjects and selected patients.
6. Learn application of techniques to vital related functions; breathing, tongue motions, facial motions, opening and closing of mouth, stimulation of soft palate.
7. Evaluate a normal subject to determine any variation in range of motion, coordination and power.
8. Outline a treatment program directed toward correction of variations or deficiencies.
9. Evaluate patients and plan treatment programs including areas of emphasis, pivots of emphasis, selection of techniques and reinforcement for patients who present flaccid paralysis, spasticity, incoordination, and orthopedic problems including postural deficiencies.

Table 1. Summary of Techniques

Procedures and techniques	Type of muscle contraction	Purposes and by-products	Indications	Contra-indications
Manual Contacts— Deep pressure but not painful. Applied to parts and muscle groups where response is desired. Manual contacts of the antagonistic pattern may be used when agonistic pattern is performed passively for determining limitations in range of motion	Isotonic or isometric	To stimulate proprioceptors in muscles and tendons and joints. May be used with or without resistance	Used whenever contact with patient by the physical therapist is necessary in an exercise procedure	Manual contacts may provide demand or security depending upon patient's needs. Manual contacts are not contraindicated except where postoperative site or open wound does not permit contact as suggested
Traction—Separation of joint surfaces by manual contact of physical therapist	Superimposed upon isotonic or isometric contractions	To stimulate proprioceptors related to stretch. To separate joint surfaces in order to make joint motion less painful	Conditions where separation of joint surfaces is desirable. When maximal facilitation is used, superimposed upon patterns the motion of which is that of pulling	Recent fractures where there is danger of separating fragments. Recent postoperative conditions if traction is generally contraindicated
Approximation— Joint compression by manual contact of physical therapist	Superimposed upon isotonic or isometric contractions	To stimulate joint proprioceptors related to compression	When maximal facilitation is used, superimposed upon patterns the motion of which is that of pushing	Same as for Traction
Stretch—Maximal stretch of major muscle components in lengthened range of pattern	Superimposed upon isotonic contractions	To demand increased response where response is inadequate to initiate active motion in lengthened range of pattern. To achieve increased response throughout the major muscle components	Conditions where innervation is inadequate to produce active motion	Acute orthopedic conditions, recent fractures, recent postoperative conditions. Pain
Timing—Sequence of contraction of major muscle components from distal to proximal	Superimposed upon isotonic or isometric contractions	To develop coordinate movements. To make possible overflow and reinforcement when resistance is superimposed	Conditions which permit active motion or motion against resistance	Only contraindicated where any form of exercise is contraindicated
Maximal Resistance —Graded according to patient's abilities and needs. May be a slight amount for weak components and a great amount for stronger components. Patient must be allowed to move if command is for	Superimposed upon isotonic or isometric contractions	To stimulate active motion. To obtain overflow from stronger components to weaker components and to reinforce weaker patterns with stronger related patterns. To develop power, endurance, coordina-	Conditions where weakness is a primary problem. Conditions which demand correction of imbalances and improvement of coordination. Conditions where relaxation is a prime need. Superimposed upon iso-	May not be superimposed upon isotonic contraction in acute orthopedic conditions. May not be superimposed upon pattern favored by an imbalance unless that pattern provides stimulation of the weaker pattern, in reversal

Table 1. Summary of Techniques (Continued)

Procedures and techniques	Type of muscle contraction	Purposes and by-products	Indications	Contra-indications
active motion. Must not be excessive so as to prevent the patient from "holding" when the command is to "hold"		tion. Correction of imbalances. To demand relaxation, to reverse adaptive shortening	metric contractions in recent fractures and acute orthopedic conditions	techniques. Must be used guardedly where sustained or prolonged effort may be harmful
Reinforcement—Accomplished by resisted motion in strongest part of range of reinforcing components of pattern. Patterns selected as reinforcement must be related and stronger than the pattern to be reinforced	Superimposed upon isotonic or isometric contraction	To stimulate weaker components or weaker patterns. To establish coordination between combinations of patterns	Conditions where weakness is a primary factor. Conditions which permit active motion against resistance	Where pattern cannot be controlled in a coordinate manner unless two hands are used on the part. Acute conditions which do not permit active motion against resistance
Repeated Contractions—Technique of emphasis. Sustained and repeated effort in one direction. May be performed at any desired point of range of motion	Isotonic following initial isometric contraction	To stimulate gains in range of active motion of agonistic pattern. To demand relaxation or lengthening reaction of antagonistic pattern. To improve endurance, coordination, and strength in a given pattern or a specific part of the range of motion of a specific pattern	Conditions where weakness, lack of endurance, and imbalance exist as primary problems	Conditions which do not permit sustained effort against resistance such as acute orthopedic and recent postoperative conditions, cerebrovascular accidents
Rhythmic initiation–Rhythm Technique—Repeated movement without sustained effort. Performed from lengthened range to shortened range	Voluntary relaxation, followed by assisted isotonic contraction, followed by resisted isotonic contraction	To promote ability to initiate movement, and to increase rate of movement	Conditions where rigidity (Parkinsonian) or spasticity prevent initiation of movement or prevent less than desirable rate	Conditions where passive movement is contraindicated
Slow Reversal—May be performed through available range or partial range according to patient's response	Isotonic of antagonistic pattern followed by isotonic of agonistic pattern. Sequence repeated if necessary to increase response	To stimulate active motion of agonistic pattern. To redevelop normal reversal of antagonists. To develop coordination of two antagonistic patterns. To develop strength in two antagonistic patterns. To achieve relaxation as a result of stimulation of the agonistic pattern	Conditions where weakness is a primary factor and where reversal provides stimulation of the agonistic pattern. Conditions have passed the acute phase and a normal reversal of antagonists is desired	Conditions where reversal does not stimulate the agonistic pattern. Acute orthopedic conditions
Slow Reversal–Hold—May be performed through available range of motion or partial	Isotonic then isometric of antagonistic pattern followed by isotonic then isomet-	Same as Slow Reversal. To develop stability and ability to perform isometric contractions in	Same as Slow Reversal. Conditions where ability to perform isometric contractions is defi-	Same as Slow Reversal

Table 1. Summary of Techniques (Continued)

Procedures and techniques	Type of muscle contraction	Purposes and by-products	Indications	Contra-indications
range according to patient's response	ric of agonistic pattern. Sequence repeated if necessary to increase response	specific patterns or specific parts of range of a pattern	cient	
Rhythmic Stabilization—May be performed at any point of available range of motion	Isometric of agonistic pattern followed by isometric of antagonistic pattern	To stimulate active motion of agonistic pattern. To develop stability of the part in specific ranges of motion. To achieve relaxation of antagonistic pattern as a result of stimulation of agonistic pattern. To stimulate circulation through isometric contraction	Conditions where weakness is a primary factor and where stabilization provides stimulation of the agonistic pattern. Conditions where active motion is not permitted or impossible because of pain. Conditions where isometric contraction is deficient as in ataxia. Stability is a goal of treatment	Conditions where stabilization does not stimulate agonistic pattern
Contract–Relax— May be performed at succeeding points of range of motion beginning with the point where limitation by antagonistic pattern presents itself	Isotonic contraction of antagonistic pattern—no range of motion allowed —followed by passive motion of agonistic pattern. Procedure followed by attempted performance of agonistic pattern from stretch stimulus or from isometric contraction in shortened range	To achieve relaxation of antagonistic pattern where active motion cannot be initiated from stretch range of agonistic pattern	Conditions where spasticity is primary factor and no active motion is available from a stretch stimulus	Conditions where active motion of the agonist is present. Acute orthopedic conditions
Hold–Relax—May be performed at any point of range where limitation presents itself as the result of pain and muscle spasm	Isometric of antagonist followed by free active motion of agonist. Isometric contraction of agonist may follow initial contraction of antagonist	To achieve relaxation of antagonist. To encourage active motion of agonist	Conditions where pain prevents active motion. Acute orthopedic conditions	Conditions where ability to perform isometric contraction is grossly deficient
Slow Reversal–Hold– Relax—Performed at exact point of range of motion where limitation by antagonistic pattern presents itself	Isotonic then isometric of antagonistic pattern— no range of motion allowed—followed by voluntary relaxation, then by isotonic of agonistic pattern. Sequence repeated in order to promote further relaxation.	To achieve relaxation of antagonistic pattern. To stimulate agonist following relaxation of antagonist	Conditions where limitation of range of motion is present and where motion against resistance is permitted. Conditions where limitation of motion is a primary factor	Conditions where Slow Reversal— Hold—Relax does not achieve relaxation of antagonistic pattern. Conditions where active motion against resistance is not permitted

Adjuncts to facilitation techniques

Certain physical agents specifically applied may enhance the patient's ability to perform (ref. 20), and at the same time, may conserve the physical therapist's energy. The agents are not new but the method of application is different. As with other techniques of facilitation, application is superimposed upon patterns of facilitation; that is, the superficial structures related to the specific patterns. The antagonistic relationship of diagonally opposite patterns and structures is considered.

If movement, active or passive, is limited by adaptive shortening, spasm or spasticity, or localized pain, the factor of limitation usually lies within the antagonistic pattern of facilitation. The relaxation or alleviation of the limiting factor may be approached through direct relaxation of the antagonistic pattern, or through the direct stimulation of the agonistic pattern with subsequent relaxation of the antagonistic pattern.

Two physical agents, cold and electrical stimulation, have been used to good advantage with the majority of patients. The use of cold has a broader variety of application and is discussed first.

COLD

Cold may be applied in several ways. One method may be selected or several methods may be combined. For direct relaxation of an antagonistic pattern which is limiting movement, turkish towels wrung from ice water are applied to the skin overlying the muscle groups of the antagonistic pattern, or in the case of painful joints, may be wrapped around the entire segment. For example, if range of the flexion–abduction–external rotation pattern of the shoulder is limited, the cold compress is placed over the axillary and pectoral regions. The compress is used for about three minutes and within this time is replaced by another cold towel at least once. With the cold compress in place, the agonistic pattern is facilitated. Relaxation techniques may be directed to the antagonistic pattern at the point in the range where limitation is evident thereby facilitating the agonistic pattern. Isometric contraction by use of rhythmic stabilization or hold–relax technique is indicated in the presence of pain. If possible, the patient should perform with isotonic contraction through the available range of motion so that a more lasting effect is obtained. If working with an extremity while the cold compress is in place seems cumbersome, related patterns of other segments may be performed while the cold is having its effect. Where pain is a dominant factor, this may be the procedure of choice.

Where a localized area of pain or limitation of movement exists, direct and specific application may be made by use of a ball of ice, formed by hand until smooth, or by use of an icicle formed on a wooden applicator (ref. 26).* The ice is rubbed vigorously over the painful area, as for example, a post-surgical scar which limits movement and produces pain. Aplication is continued until the patient no longer "feels" the cold, usually less than one minute. Positioning and maintaining the part so that tension is present in the limiting muscle groups or soft structures is conducive to maximum relaxation providing that pain is not produced. If tension produces pain, the degree of tension should be reduced. In this way, the movement–pain–limitation cycle may be interrupted. The point of pain may have been altered. That is, movement may continue to produce pain, but the range of pain-free movement has been increased. When resistance is superimposed, it is applied most strongly to the pain-free components of a pattern; if movement of the proximal joint is painful, maximal resistance may be applied to distal muscle groups during performance of isometric and isotonic contractions. Again, the use of cold is coupled with exercise so as to utilize any relaxation that has been gained.

Selective stimulation or facilitation of a specific

* Margaret Rood has used plastic trays commercially made for home use in preparing "Popsicles" for children. Miss Rood has also extended the use of cold for discrete sensory stimulation and inhibition.

muscle or muscle group may be gained by discrete application of cold (ref. 26). Using an ice ball or an icicle, a quick stroking of the skin overlying the muscles of the agonistic pattern may promote response of these muscles. If, for example, flexion–abduction–external rotation of the shoulder is painful, quickly and briefly stroking the areas of the trapezius muscle and the middle portion of the deltoid muscle may increase the pain-free range.

Immersion of a segment in ice water may be useful for relaxation of the distal musculature. The hand and forearm or the foot and leg may be immersed for a minute or less and increased as tolerated. If tolerance for immersion is low, dipping the part briefly and repeating the process several times may be desirable at first. The distal part of the segment will be relaxed. Should limitation persist at the proximal joints, cold compresses may be used proximally. Upon completing the procedure, exercises and facilitation of the desired movements should be carried out.

Immersion of the lower region of the body and extremities in a cold bath, about 50°F., for one to four minutes may help to reduce marked spasticity in patients having generalized involvement. The procedure should be adapted to the patient's tolerance. Mead is of the opinion that the contraindications for the use of cold are rare (ref. 23). However, the individual patient must be assessed by his physician, and any possible contraindications must be heeded. Patients are not subjected to sudden application of cold without preparatory discussion. Physical therapists and physicians should undergo the treatment procedure so as to understand better patients' reactions. In general, patients like cold, although in the beginning, some are not completely receptive to the idea of its use.

Cold is used as preparation for exercise and movement, and for relief of pain experienced during movement. Thus cold is used locally on a treatment table, on a gymnasium mat, or in a gait area. If necessary, mat and gait activities are carried out in the privacy of a treatment room so that the part to be treated may be exposed.

ELECTRICAL STIMULATION

The use of faradic or tetanizing current for relaxation of spasticity or of adaptive shortening is a useful preliminary to performance of patterns of facilitation (refs. 20, 21). As a procedure, electrical stimulation is more time consuming than is the use of cold. However, in selected patients, and where the use of cold is medically contraindicated, electrical stimulation may be preferred. Proper application requires that two physical therapists work together; one person controls the stimulation, the second moves the segment of the patient's body passively through the available range of motion as relaxation occurs.

For electrical stimulation, the usual preparations are carried out. The thoroughly moistened inactive electrode (about 3 inches by 4 inches) is placed at a distance from the part to be stimulated. That is, if the lower extremity is to be stimulated, the inactive electrode is placed near the mid-thoracic region so that contact will be maintained as the extremity is moved. The active electrode (about 1 inch in diameter) attached to a long-handled applicator is used for ease of control and application. The active electrode is applied to the skin with firm pressure. After the electrode is in place, the current is increased sufficiently to produce a tetanized contraction. The current is decreased before the electrode is withdrawn. In this way, the patient having intact sensation experiences less discomfort.

The therapist who is responsible for moving the segment determines the areas of limitation by passively moving the part through the available range of motion. The movement is carried out by moving the distal parts first, then progressing to the more proximal joints. If the patient is able to move the part actively or to assist he is instructed to do so. The areas of limitation or points at which limitation becomes evident having been determined, the stimulation is done in a proximal to distal direction. The proximal muscles of a pattern are stimulated first; strict adherence to classical motor points is not essential. Application of the active electrode to skin overlying those muscles which are diagonally opposite the limiting pattern is important. For example, if the biceps femoris muscle is limiting complete extension of the knee, the vastus medialis muscle is stimulated. The therapist who is supporting and moving the segment waits for relaxation to occur. As tension lessens, the part is moved through additional range. A degree of relaxation having been achieved, the active electrode is moved to a somewhat more distal point over the same muscle or over a related muscle of the same pattern. The entire procedure is repeated and the distal muscles of the pattern are stimulated successively. The person who is moving the part is able to direct the person who is controlling the current because of his awareness of exact points of limitation. Because overlapping exists between the patterns of one diagonal and those of the second diagonal, it may be necessary to relax the antagonists of the second diagonal as well.

This is not done in a haphazard fashion. Again, all components of the second diagonal are considered and the procedure is conducted from proximal to distal.

After stimulation has been completed, the passive and active range of motion may be tested to determine the degree of success. In any case, the agonistic pattern should be facilitated as soon as possible by use of maximal resistance. Repeated contractions should be performed so as to promote a more lasting effect. If possible, the patient should perform actively the movement or activity in whose interest the procedure was done. It follows that the greater the potential for performance actively and against resistance, the greater the lasting effect.

3. Facilitation of total patterns

Related aspects of motor behavior

The development of motor behavior is expressed in patterned movement. An ordered sequence of motor acts emerges in the normal process of growth. The overt manifestations of growth and development bear analysis. Hooker (ref. 15) observed early fetal activity and Humphrey (ref. 16) has identified the corresponding anatomical patterns. Gesell and his coworkers (ref. 6), and McGraw (ref. 22) recorded their observations of the ever-changing and interweaving of activities as motor behavior grows and matures after birth. The observations of these workers are, in a sense, "clinical" observations. That which can be seen has been noted. The "total structure in action" can be seen; developing patterns of movement can be observed. The component patterns of total movement patterns can be analyzed.

TOTAL TO INDIVIDUATED

Motor development has certain elements and characteristics that have been identified. Hooker (ref. 15) in his studies of early fetal activity, found that responses occur first to sensory stimulation around the mouth and that the response is a total one; all segments which are functioning participate in a mass movement. The head and neck flex lateralward. This is followed by lateral flexion of the trunk with extension of the arms. Later in the developmental process, an individual segment may be stimulated and will respond specifically without a total, mass response. Fetal movements are reflex in nature and may be termed primitive for the human species. However, they are the forerunners of purposeful movement.

PROXIMAL–DISTAL TO DISTAL–PROXIMAL

The development of motor function in the fetus (ref. 15) has direction. The direction is from head to foot, cephalo-caudad, or from superior to inferior regions of the body. The direction is also from proximal to distal; that is, movements of the neck and shoulders occur before movements of the hand are evident. Gesell has simply stated, "The or-

ganization of behavior begins long before birth; and the general direction of this organization is from head to foot, from proximal to distal segments. Lips and tongue lead, eye muscles follow, then neck, shoulder, arms, hands, fingers, trunk, legs, feet" (ref. 6). Sensory development, too, is cephalo-caudal; but when sensation has arrived in the feet and hands, stimulation of a segment produces a sequence of movement in a distal to proximal direction (ref. 15), that is, when the palm is stimulated, the fingers flex and the wrist flexes. Such is the beginning of timed, coordinated movement.

REFLEXIVE TO DELIBERATE

After birth, while the developmental process continues in a cephalo-caudad and proximal-distal direction, the first movements and postural attitudes as observed by McGraw (ref. 22), and Gesell and his co-workers (ref. 6) are reflex in character. The response to a startling stimulus (Moro reflex) is a total response of bodily movement. The asymmetric tonic neck reflex induces a postural attitude of the total structure. These reflexes have components used later in rolling from supine to prone. In the newborn, the turning of the head, the roving movements of the eyes, grasping of the fingers, rapid flexion and extension of the lower extremities, and stepping movements are reflex responses ready for transition to functional movements. As the growth process continues and the repertoire of activities enlarges, movements take on an automatic quality; the child appears to practice a newly acquired movement (ref. 22), as, for example, while rolling from supine to prone. He repeats rolling well in advance of using the movement to assume a sitting posture. As he progresses, his rolling movements become more deliberate and he incorporates them in functional activities. He uses rolling from supine as he arises to a sitting position. He rolls from supine to prone as preparation for progression in the prone position. In the entire course of development of motor behavior, primitive responses give way to controlled movements and postures which may be achieved automatically or deliberately as the occasion requires (ref. 22).

MOTILE TO STABILE

Another characteristic of developing motor behavior is that movement precedes sustained posture. When the fetus is stimulated, resultant movement fades and terminates. After birth, motility is a striking feature of a newborn's behavior (ref. 22). The newborn infant moves his extremities rapidly, but unless he cries these movements are rarely seen as a sustained effort. The postural and righting reflexes are invoked by movement, by alterations of the position of the head in space and in relation to the trunk and extremities, or the trunk and extremities in relation to the head (ref. 24). As with other aspects of developing motor behavior, the righting response is composed of reflexes that have developed in a cephalo-caudad direction (ref. 22). From quiescent positions, such as supine, lateral (sidelying), or prone, movement is necessary to alter position. In this respect, movement may be considered more primitive than sustained posture. Yet, as motor behavior matures, the stability of sustained posture is necessary for purposeful movement.

OVERLAPPING TO INTEGRATIVE

Motor abilities develop in sequence and those which appear early in the sequence overlap with or contribute to those which emerge later. This characteristic of development may be observed in the normal child. There is an interweaving of component movement patterns and component postures. One activity prepares the way for another. For example, rolling is a component of the human righting response, the achievement of an erect posture (ref. 22). Thus, the ability to roll leads to the ability to assume and sustain a sitting posture. The ability to assume and sustain a sitting posture leads to the ability to assume and sustain a standing posture. The ability to roll from supine to prone and from prone to supine prepares for the ability to creep. The ability to roll and the ability to creep lead to the ability to walk. The entire process is in continuum. Within the sequence of interrelated activities, motor behavior becomes integrated and movement becomes coordinated, functional, selective, and versatile.

GROSS TO SELECTIVE

While the development of component movement patterns interweaves and overlaps, the participation of bodily segments in relation to the neck and trunk becomes more controlled, more varied as to degree and range of movement and, therefore, more complex. At first, movement tends toward full ranges —complete range of flexion, then complete range of extension. There is an oscillation between extremes (ref. 22). Then later, as integration occurs, with controlled posture and movement interacting as necessary, direction and range of movement are subservient to the total activity or pattern of movement.

A total pattern of purposeful movement, such as walking, has direction which may be continued, discontinued, or reversed at will. Reversal of movement occurs within a total pattern, i.e., the reciprocation of extremities during walking. While the total pattern of walking has a forward movement, the forward direction is achieved through reversing movements. That is, there is an alternation of activity between opposite component patterns of movement as in dorsiflexion, then plantar flexion of the foot and ankle. Thus, alternation of activity, reversal of movement, occurs between component patterns of movement and within component patterns; and the total pattern may be reversed.

By the time motor ability has matured, innumerable combinations of movement of head and neck, trunk, upper and lower extremities may be performed. There is selectivity and combining of movement patterns.

The extremities contribute component patterns to a total pattern in various ways. At first, in the supine position, movements of the upper and lower extremities tend to be symmetric (like movements of upper or lower extremities at the same time), although asymmetric (movement of upper or lower extremities to one side at the same time) and alternating movements do occur (ref. 6). As the child learns to roll, ipsilateral movements (arm and leg of same side) participate. In the prone position, ipsilateral or symmetric movements occur as well as alternating movements of arms and legs (ref. 22). As prone progression is achieved, simultaneous movements of an upper extremity and the contralateral (of the opposite side) lower extremity contribute reciprocally to crawling and creeping.

In fetal activity and in newborn behavior, flexion truly dominates extension (ref. 22). Although flexion and extension continue to be the major components of motion, combinations of these with the components of adduction and abduction, external rotation and internal rotation become increasingly apparent as the child's repertoire of movements enlarges. He is not limited to one direction of movement. He utilizes combined components as he moves forward, backward, turns in a circle, moves sideward, and diagonally in all directions. All component patterns of motion within the activity, the total movement pattern, serve his purposes in varying degrees at various times as he needs them. In the sitting posture, the child is free to move his upper extremities bilaterally in symmetric and asymmetric movements or in unilateral (one arm, or one leg, at a time) movements, or in a reciprocal (opposite movements of opposite extremities at the same time) fashion. In walking, he uses reciprocal motions of upper and lower extremities. He jumps with bilateral movements. Hopping is a unilateral activity for one lower extremity. Skipping and leaping require reciprocation of the extremities.

INCOORDINATE TO COORDINATE

During the time that the infant is developing his neuromotor ability, maturation of his special senses, too, is progressing (ref. 22). Developing vision serves movement and movement serves vision. Hand-eye coordination in reaching and prehension are sensori-motor activities certain foundations for which were laid down in the tonic neck reflex pattern (ref. 6). The developing child looks or gazes at the object for which he reaches and grasps. Movements of the hand and arm (and of the head, neck, trunk, and other segments as the occasion demands) may follow the visual phase of moving the eyes and looking, and, in turn, the visual act may follow movement of the hand. When an object is within reach, the child may regard it either before or after he grasps it. Locomotor activities and prone postures where the hands and arms reach or support weight must contribute to the development of prehension and manipulative skills. When the infant is prone, his fingers extend as he supports his weight on his palms with elbows extended. When he releases support on his arms, he reaches an arm forward and may flex his fingers and pull on the surface (ref. 22). This type of finger extension and flexion precedes and promotes functional grasp and release of objects. The development of auditory responses, too, plays an important role in the development of motor behavior. As the infant begins to locate the sources of sounds, he turns his head appropriately (ref. 6). Sound becomes a stimulus for movement. As he interprets sound and language, he begins to move in response to sounds and words.

COORDINATED MOVEMENT

The striking feature of *mature* movement is that it is coordinated. Other features of strength, endurance, and rate of movement that support coordina-

tion are evident well in advance of purposeful and functional movement. That is, the infant exhibits strength as he grasps and of the total musculature during crying. He exhibits endurance for repeating movement. He moves his extremities rhythmically and rapidly or slowly, although in random fashion. However, except for movements which have to do with vital activities, such as breathing, sucking, swallowing, peristalsis, micturition, and defecation, the newborn's movements lack a purposeful quality. As he grows, as motor behavior becomes organized from head to foot, the coordination of his movements, too, proceeds from head to foot. The upper extremities reveal coordinated movement, as in precise prehension, before the lower extremities are fully developed for independent walking (ref. 6).

TIMED INTERACTION

In order for a movement to be coordinated it must have a sequence of action and interaction within and between segments. The sequence and timing of a movement permits ease of movement with economy of effort. There is a neat balance and counteraction between antagonistic components. For example, in rolling from supine to prone toward the left, the head turns toward the left and the neck proceeds into extension. As the head and neck move and the spine extends, the right upper extremity is prepared to lead, while the foot of the right lower extremity may push against the surface so as to elevate the pelvis; then the right upper and lower extremities lift across toward the chin side (to the left). Ordinarily the upper extremity leads the rolling movement, but variations on this sequence may occur. Later, the lower extremity may be lifted first and then the upper extremity may follow (ref. 22). Nevertheless, the act is coordinated and exhibits a timed sequence of component patterns of movement of the extremities in relation to the head, neck, and trunk. As the timed sequence is established, the movement is reproducible on demand.

In coordinated, mature movement the sequence within the extremities is from distal to proximal. Again, in the act of rolling, the leading hand and arm move before the shoulder girdle is completely elevated. If the shoulder were to move first, the movement of the hand and arm would appear to be an "afterthought." If the pelvis were to be completely elevated and rotated before the foot pushed against the surface or proceeded to lift across, the lower extremity would deter rolling rather than enhance it.

SEQUENTIAL ACTION OF MUSCLES

Coordinate movement with body parts interacting in sequence includes a sequential action of muscles.

Those muscles necessary to the accomplishment of a movement will respond in the necessary sequence, and in a distal to proximal direction, although the development of the musculature has been from proximal to distal (ref. 7). As in rolling to prone, when the foot pushes on the surface, plantar flexion with eversion serves through action of the toe flexors, the peroneus longus, and the gastrocnemius. Having pushed, the foot then serves the lifting of the extremity by dorsiflexing with inversion through action of the toe extensors and the anterior tibial, since the lifting phase is toward the midline of the body. However, if the lower extremity is lifted without push-off, it is lifted with the foot dorsiflexing, and there is no need for plantar flexion with eversion of the foot. Plantar flexion without push-off would deter the forward movement of the lower extremity by introducing an antagonistic movement; it would disturb the sequence of the total pattern of muscle action, and thus the coordination of the movement.

SUMMARY

In summary, the development of motor behavior is expressed in an orderly sequence of patterned movements. Growth of sensori-motor activity has a cephalo-caudad and proximal-distal direction, but coordinated movement has a distal to proximal sequence. Movements which are primitive and reflex in nature are altered by growth so that movement becomes automatic and then deliberate or purposeful. Upon maturity, coordinated, functional movement may have both automatic and deliberate aspects. Movement precedes postural control; movement is necessary to alter position or posture; posture is necessary to purposeful movement. In the sequence of developmental activities, the component patterns of movement and posture which make up one activity contribute to or overlap with the components of other activities in the sequence. Sensory development and motor development progress together and are inseparable. The end result of the process is a vast repertoire of coordinated movements and combinations of movements.

Rationale

The fundamental motor activities of the developmental sequence are interrelated and universal. Every human being who is capable of normal movement and balanced postures has learned to roll from supine to prone and from prone to supine, to progress or move in the prone position, to assume a sitting position, to arise to an erect posture, and to walk, run, jump, hop, skip, and leap. Individual variations occur in regard to mode of performance and sometimes as to sequence, for according to McGraw, the impulse to progress is "somewhat specific and may be 'grafted onto' any postural form predominant at the time the urge to progress is manifest" (ref. 22).

By the time the child has matured in his motor activities, he can perform in a coordinate fashion all activities in the sequence. Primitive patterns of movement have been altered by growth; mature movements of a willed or voluntary nature have become dominant but they retain automatic and reflex aspects. The normal adult may revert to more primitive responses in stress situations. If when lying on the beach a person senses impending danger (approaching flames from a fire, a snake passing by, a man with a dull instrument raised overhead), he may automatically roll away from the hazard. The act of rolling serves his immediate need best. He may then from a prone position, or by rolling to sitting, scramble to his feet and walk or run away as the situation demands. He has used automatically a sequence of developmental activities which had their origin before birth, which matured within the first few years of life, and which may not have been used in sequence for a number of years, but which were immediately available on demand.

The ingredients from which normal movement is made, the emergence of specific patterns of movement from total patterns of response, the primitive and reflex aspects which underlie controlled posture and movement, the direction of development from head to foot, the distal to proximal direction of coordinate movement, the refinement of movement from full range to partial and specific ranges, are all characteristics of the developmental process which provide a basis for development or restoration of motor function in persons who lack normal ability to move or to sustain posture (ref. 17). For these persons, a recapitulation of the developmental sequence is a means to the end—the ability to care for one's body, to walk, and to engage in productive work.

Principles

Certain principles guide the use of developmental activities when patterns and techniques of proprioceptive neuromuscular facilitation are superimposed.

1. Developmental activities are useful as a basis for treatment of patients of all ages. The chronological age and the level of development must be considered. Aging is a normal process of human development in which alterations occur in postural form and configuration of movement (ref. 3).

2. The reflex mechanisms underlying normal movement are recognized as potent forces for influencing movement and posture (ref. 13). The sequence of developmental activities, by virtue of the normal process of growth and development, provides for activation of postural and righting reflexes. The coordination of visual-motor mechanisms and of auditory-motor mechanisms are taught and are utilized in training.

3. Development or restoration of motor abilities, including self-care and the ability to walk, are concomitants of motor learning. Patterns and techniques of proprioceptive neuromuscular facilitation are used to hasten motor learning by providing appropriate "sensory cues." The selection of sensory cues is the task of the physical therapist or teacher (ref. 9).

4. Repetition of coordinate movement is used to increase strength and endurance and to adjust the rate of movement (ref. 11). Resistance is applied during repetitions but is graded according to the needs and abilities of the patient.

5. In the developmental process, the development of movement is from proximal to distal and total patterns of movement become individuated (ref. 15). In using the developmental sequence with emphasis first on training of head and neck and trunk patterns, the proximal to distal relationship and the progression from total patterns of movement to individual component patterns are heeded.

6. That coordinate movement proceeds in a distal to proximal direction is recognized as essential to development or improvement of motor abilities (ref. 15). In applying patterns and techniques of proprioceptive neuromuscular facilitation, a timed sequence of movement from distal to proximal is used.

7. Developmental activities are total patterns of movement and posture to which patterns and techniques of proprioceptive neuromuscular facilitation are applied precisely. The component patterns of a total pattern are readily converted to spiral and diagonal patterns of facilitation for maximum selectivity of response. Techniques based on isotonic contraction of muscle encourage movement; those based on isometric contraction encourage stability and sustenance of posture.

8. For optimum development of motor function, the patient must be helped to recapitulate the developmental sequence insofar as he possibly can. Each phase of the sequence has significance in that one activity lays the foundation for a more advanced activity. If a phase is omitted, function may be altered unfavorably and certain deficiencies may be retained unnecessarily. The acts of assumption of positions are necessary to maintaining balance in the positions.

9. The stronger component patterns of a total pattern, and the stronger pivots of action within a component pattern, are utilized for augmentation of weaker components. By using the patient's abilities to lessen his inabilities, an activity within the developmental sequence may be learned more readily.

10. The progress of the patient is enhanced by adequate performance of an activity within the sequence rather than by inadequate performance of a variety of activities. Performance of more primitive activities, insofar as is possible, should be instructed before attempting more complex activities that are completely beyond the ability of the patient.

11. The physical therapist becomes a part of the total movement or effort of the patient. The therapist must approach the patient in a mutually advantageous way. As patterns of movement have a diagonal direction, the therapist assumes and moves in a

Facilitation of total patterns

diagonal direction as the patient moves. This principle applies wherever activities are performed—on a gymnasium mat, a bed, a treatment table, or in an open area during gait training.

12. The program of activities is selected in accordance with the patient's needs and potentials. Long-term and short-term goals must be determined.

All activities must be integrated and directed toward suitable goals.

Thus, the total approach of proprioceptive neuromuscular facilitation has denominations of total patterns of movement, specific patterns of facilitation combined for training of total patterns, and techniques for hastening motor learning.

Use of the developmental sequence

The developmental sequence is limited to those activities which are most typical of *human* development. McGraw's ref. 22) reminder that man as a species has a long history of evolution serves as a guide to selection of primitive activities. Every person concerned with development or restoration of motor ability in others can readily learn and apply the sequence.

As selected and adapted from McGraw (ref. 22) and Gesell (ref. 6), the developmental sequence provides for a progression from primitive movements and postures to more complex and advanced movements and postures. Briefly, the sequence of total patterns of movement with their related or ultimate positions and postures proceeds as follows: rolling from supine to prone, and from prone to supine; prone progression, as in pivoting, crawling, creeping, and plantigrade walking; rising to sitting; rising to kneeling, and knee walking; rising to standing, and bipedal walking; ascending and descending stairs and ramps; running; jumping; hopping; and skipping. Table 2, Progression of Activities, portrays the progression of the elementary activities in the sequence. Within the sequence, movement is used to alter position and posture. Movement is enhanced by use of the hands following the eyes, or the eyes following the hand movements. Movement and posture, including eye-hand co-ordination, become interwoven. The sequence provides for total patterns wherein the head and neck, the trunk, and the four extremities participate in various relationships such as ipsilateral, bilateral symmetrical, bilateral asymmetrical, and reciprocal movements. Within a total pattern, certain segments may move while other segments adjust to the movement.

In the sequence, as position is altered by movement balance and posture are developed and maintained in the altered position before movement leads to another position. That is, when the hand and knee (creeping) position is achieved, balance in this position precedes the creeping movement. Balance

in a position is not limited to neutral positions of the extremities; positioning of the extremities in various ranges of motion and for reciprocation is used.

Provision is made for flexor dominance of a movement and for extensor dominance. As an illustration, the sitting position is assumed by initial rolling from supine toward prone (flexor dominance), and then by pushing up from the prone position (extensor dominance). Also, when balance is to be maintained in a certain position, the flexor components are required to alternate with extensor components. This alternation occurs when balance is disturbed in an anterior to posterior direction and then in a posterior to anterior direction. For instance, when a patient attempts to maintain himself in a lateral (side-lying) position, flexors dominate if balance is disturbed toward the supine position, and extensors dominate if balance is disturbed toward the prone position. In prone progression, such as in creeping, dominance is altered by changing direction from forward (flexor dominance) to backward (extensor dominance). As identified in Table 2, dominance is influenced by direction and by performance against resistance.

The developmental sequence promotes the ability to perform with isotonic contractions of related muscle groups during movement and with isometric contractions during performance of balancing activities. Transition from isometric contraction to isotonic contraction also occurs. Sherrington (ref. 27) pointed out that, "Naturally, the distinction between reflexes of attitude (posture) and reflexes of movement is not in all cases sharp and abrupt. Between a short lasting attitude and a slowly progressing movement the difference is hardly more than one of degree. Moreover each posture is introduced by a movement of assumption, and after each departure from the posture, if it is resumed, it is reverted to by a movement of compensation. Hence the taxis of attitude must involve not only static reactions of tonic maintenance of contraction, but innervations which execute reinforcing movements and

Facilitation of total patterns

compensatory movements." If balance is disturbed so that the position must be recovered, recovery is achieved through isotonic contraction. During movement, stability of supporting segments must be maintained through isometric contractions as one segment advances with isotonic contractions of the responsible muscle groups. However, transitions between and intermingling of types of muscle contraction are not sharply distinct.

Ultimately, the use of developmental activities contributes to independence in self-care and gait activities. As an example, rolling movements are closely related to turning in bed, rising and sitting on the edge of the bed, dressing while supine, and when the component movements of the upper extremity are considered, to feeding and other activities requiring hand-to-face movements. Locomotion in the prone position and rising to standing prepare for bipedal walking. A patient's potential for performance may be limited by pathology but insofar as is possible the performance of the developmental sequence should promote optimum recovery of purposeful movement.

MAT ACTIVITIES

Practical application of the principles for use of developmental activities requires proper physical facilities. A gymnasium mat should be firm, smooth, and pliant enough so as to be comfortable and to protect the patient from abrasions and undue stress should he fail to maintain his balance. An uninterrupted surface, without seams or tufts, which can be easily cleansed, seems best.

A mat's surface should be large enough to accommodate the patient and the therapist for creeping and walking activities with several repetitions of movement before direction is changed. For an adult, a useful size is six feet by eight feet; for a child, four feet by six feet. Elevation on a platform the height of the seat of a standard wheel chair is desirable so that transferring from chair to mat may be incorporated as a related self-care activity. If a mat is used on the floor, a small ramp with hand rails allows some patients to descend from chair to mat and then return to the chair with minimal guidance and protection. For working with an individual patient, one mat of suitable size is sufficient. However, a series of mats covering an area of at least twenty-four feet by twenty-four feet will permit a number of patients to be supervised in those activities which they can perform independently.

Motor activity on a gymnasium mat has certain advantages. There is an element of security for patients who are fearful of falling. Total patterns of movement may be performed without limitation such as exists when the patient is treated on a table. Postural and righting reflexes may be induced more effectively because a variety of positions and postures may be used, and because balance may be disturbed without the hazard of falling from a height. In-bed activities may be simulated and practiced safely as a means of teaching self-care, inasmuch as the sequence provides for a close relationship with self-care activities. Furthermore, when mat activities are done in an open area where patients can observe each other, there are indirect influences, such as learning from others, motivation through competition, and socialization, all of which may enhance performance.

Precise use of patterns of facilitation and application of techniques must be in keeping with the abilities and needs of the individual patient. While the progression of activities is outlined in terms of total patterns of movement and related positions of balanced posture, precise use demands further analysis of a total movement or posture. The component patterns within the total movement or posture must be identified throughout the neck, trunk, and extremities. By so doing, the components of motion within a component pattern can be utilized for augmentation and reinforcement of the patient's effort to perform the total pattern, since all movement proceeds in the same direction.

A total pattern of movement and each component pattern have a point of initiation, a range of movement, and a point of completion. The point of initiation is termed the lengthened range; midway within the range of movement is the middle range; and the point of completion is the shortened range. A total movement, or one of its component patterns, may be initiated in the lengthened range with isotonic contraction of responsible muscle groups. Movement may be initiated in the shortened range or the middle range with isometric contraction followed by repeated isotonic contractions. If movement is initiated in the lengthened range the direction proceeds toward the middle or, if possible, the shortened range. If movement is initiated in the middle or shortened ranges, the direction is still toward the shortened range, the point of completion. In this instance, however, repetitions of effort are controlled in retrograde so as to increase the range through which a movement is performed. This becomes necessary to the development of the complete range of a movement; the technique of repeated contractions is used.

Table 2. Progression of Activities

TOTAL PATTERN OF MOVEMENT	POSITION FOR BALANCE

Rolling from supine to prone
Flexor dominance (F.D.)

Lateral (sidelying)

Rolling from prone to supine
Extensor dominance (E.D.)

Lateral (sidelying)

Prone pivoting
Alternate F.D. and E.D.

Prone on elbows and pelvis

Rising to elbows and knees from prone,
E.D.

Prone on elbows and knees

Pulling to sitting from supine,
F.D.

Sitting with hands reaching forward

Rising to sitting from hyperflexion,
E.D.

Sitting with hands supporting forward

Crawling on elbows and knees
Forward, F.D.; backward, E.D.

Prone with elbows and knees in various positions

Rising to hands and knees from prone,
E.D.

Prone with hands and knees in various positions

Rocking on hands and knees
Forward, E.D.; backward, F.D.

Prone on hands and knees while rocked forward, and backward

Creeping on hands and knees
Forward, F.D.; backward, E.D.

Prone on hands, one knee and one foot

Rising to sitting from prone
E.D.

Sitting with and without hands supporting backward

Rising to sitting from supine
F.D.

Sitting with and without hands supporting forward

Rising to plantigrade position from prone,
E.D.

Plantigrade, hands and feet in alternate ipsilateral

Plantigrade walking,
Forward, F.D.; backward, E.D.

Plantigrade with diagonal reciprocation, and with one extremity free of support

Pulling to upright from hands and knees
(stall bars) F.D.

Upright for climbing with hands grasping and with support on one knee and one foot

Climbing and descending (stall bars)
ascent, F.D.; descent, E.D.

Climbing with reciprocation

Rising to kneeling from sitting
(buttocks on heels) F.D.

Kneeling on both knees

Facilitation of total patterns

Table 2. Progression of Activities (Continued)

TOTAL PATTERN OF MOVEMENT	POSITION FOR BALANCE

Rising to kneeling from prone
E.D.

    Half-kneeling, one knee and one foot

Walking on knees
Forward, F.D.; backward, E.D.

    Kneeling with diagonal reciprocation

Pulling to standing from squat or sitting,
F.D.

    Standing with hands and feet parallel positions

Rocking on feet with hands supported
Forward, E.D.; backward, F.D.

    Standing in ipsilateral position with and without hands supported

Lowering to squat or to sitting
F.D.

    Near-squat and near-sitting with and without support of hands

Rising to standing from prone
E.D.

    Standing with reciprocation with and without hands supported

Rising to standing from supine
F.D.

    Standing with reciprocation with hands supported and with one foot free

Foot lifting and stamping
Alternate F.D., E.D.

    Standing with reciprocation with one hand free and one foot free

Bipedal walking
Forward, F.D.; backward, E.D.

    Standing with reciprocation with both hands and one foot free, and on tiptoe

Ascending and descending ramp on hands and feet,
forward ascent, F.D.; backward descent, E.D.

    Plantigrade with reciprocation, and with one extremity free

Ascending and descending ramp in upright, forward
ascent, F.D.; forward descent, F.D.; backward
ascent, E.D.; backward descent, E.D.

    Standing with both hands supported, one hand free, one hand and foot free, hands free and one foot free

Climbing and descending stairs on hands and feet,
forward ascent, F.D.; backward descent, E.D.

    Plantigrade with bilateral position and reciprocation

Ascending stairs in upright
(advancing one foot; advancing alternately)
forward ascent, F.D.; backward descent, E.D.

    Standing with both hands and feet supported, one hand free, and hands free

Descending stairs in upright
(advancing one foot; advancing alternately)
forward descent, F.D.; backward ascent, E.D.

    Standing with both hands and feet supported, one hand free, one hand and foot free, hands free

Running, jumping, hopping and skipping

Note: Chronology of the sequence has been omitted. The sequence is used in accordance with a patient's needs; chronological age is one factor of evaluation and treatment.

Solid lines indicate progression from assumption of a position to balance or posture related to the position, and from balance in one position to a more advanced form of movement or locomotion. Interrupted lines show relationships within prone postures and locomotion.

Dominance is influenced by direction and by superimposing manual resistance.

Use of the developmental sequence

A total movement may take several directions, such as forward, backward, sideward, or turning in a circle. As direction is changed, the component patterns and components of motion change accordingly. Flexor dominance of a movement may be altered to extensor dominance by changing direction from forward to backward. Lateralward (abduction) movements in a sideward direction may be altered to medialward (adduction) movements by changing direction to the opposite side. Turning in a circle and changing direction provides for lateralward and medialward movements of contralateral upper and lower extremities. Diagonal movements combine the flexion or extension components with abduction and adduction components. Diagonal movements permit greater selectivity than is possible with straight forward or straight backward movement.

The physical therapist approaches the patient so as to be in line with the movement of the patient. Techniques of proprioceptive neuromuscular facilitation may be applied to each total pattern of movement. Through maximal resistance, stronger component patterns will augment the response of weaker component patterns. When the goal is to *move,* movement must be permitted; but the stronger patterns must be resisted strongly. To reinforce a weaker component pattern by a stronger component pattern, the technique of timing for emphasis is used. The patient is instructed to move; as he moves he is asked to sustain his effort ("hold" with isometric contraction) at the strongest phase, and then repeated isotonic contractions of the weaker pattern are demanded and commanded.

When the goal is to *maintain balance* in a certain position or posture, a stronger segment, or pattern within a segment, is again used to augment the ability of a weaker segment to sustain posture. The patient is instructed to maintain the position and as he does so, the technique of rhythmic stabilization is used along with approximation applied to the weight-bearing segments. Resistance is applied first to the isometric contractions, holding, of stronger patterns or segment, and then an isometric contraction is gradually developed in the weaker patterns or segment. Counterpressure with anterior and posterior manual contacts applied simultaneously promote maximum stability and security. If balance is to be disturbed sufficiently to promote compensatory movements, both manual contacts may be anterior, or posterior, as necessary. Compensatory movements are resisted. Balance may be disturbed abruptly by brief and suddenly applied and withdrawn pressure. Manual contacts must be

selected so that the desired patterns will be evoked. When balance is disturbed in this fashion, the response is quick, and resistance is not used.

While the various techniques may be applied, the use of timing, reinforcement, maximal resistance, traction and approximation, stretch, reversal of antagonists including rhythmic stabilization, and repeated contractions seem most useful. When the ability to move is deficient, techniques which require isotonic contractions of muscles should be selected. When stability is deficient, isometric contractions should be used. As always, due regard must be given the prevention and correction of imbalances between antagonistic reflexes, patterns of facilitation, muscle groups, and components of motion.

ILLUSTRATIONS

The illustrations (Figs. 38–75) with their legends portray the use of selected activities from the developmental sequence. However, the sequence of the illustrations is not developmental. Rather, closely related activities appear in series; rolling activities are grouped together (Fig. 38–48), as are the activities related to prone progression (Figs. 51–63). Table 2, Progression of Activities, should be used as a guideline for appropriate overlapping between activities in various postures. Descriptions of standing balance and walking (pages 162 and 168–169) are presented as gait activities and should be read in conjunction with the related illustrations for mat activities.

To portray adequately all activities, all variations of posture, and all variations on the therapist's approach to the patient for performance to left and right, would require hundreds of illustrations. This series is a sample. Some of the significant aspects of mat, gait, transfer, and self-care activities to be observed are:

1. the normal subject is shown with bare feet so as to reveal responses of the feet more clearly (during mat activities neither patient nor therapist wears shoes. In gait and transfer activities shoes are worn);
2. movements of assumption of posture, the posture assumed, and the mode of progression related to the posture;
3. the coordination, or combining, of component patterns of a total pattern wherein vision, head, and neck lead and give direction;
4. the component patterns to which resistance or a technique is directed, the component patterns of segments which may move free of manual contact, but not necessarily free of resistance,

or which may adjust with a compensatory movement as balance is disturbed;

5. three phases of a total pattern of movement, or of a component pattern: lengthened, middle, and shortened ranges (in some instances, the initial position is used, and in others, where the range is incomplete, indication as "approaching" middle or shortened range is used); or three variations on a total pattern of posture during balancing activities;

6. the therapist's approach to the subject (or patient) by taking a position that is on the diagonal if diagonal direction is expected, and by assuming a total posture that will permit movement to occur or posture to become stable, whichever is indicated;

7. a variety of manual contacts, but not all-inclusive, that must be transposed for performance to the opposite side which also requires that the therapist assume a transposed or adjusted posture;

8. commands that include some preparatory instructions as well as commands to perform, and in some instances are in keeping with the suggested techniques (commands for a specific technique should be gleaned from the text on the technique; inclusion here is only a guideline);

9. suggested techniques which may be superimposed (many other techniques may be considered for use. Those suggested here should be treated as suggestions for learning);

10. antagonistic patterns, some of which are illustrated, some of which are not so as to permit portrayal of a variety of combinations of component patterns and of manual contacts (performance of an antagonistic pattern requires adjustment of the therapist's position and manual contacts, and where equipment is used, for example, a wheel chair, alteration of the position of the chair may be necessary);

11. augmentation, reinforcement of a total pattern of movement or posture by use of component patterns; and the use of a total pattern for augmentation of a component pattern (reinforcement by postural and righting reflexes is evident in many of the illustrations);

12. notes which include comments on the therapist's and the subject's performance, additional suggestions for performance, and in some instances references to preparatory and related activities.

Within the sequence of activities as portrayed in Table 2, there is an overlapping between mat and gait activities. The most advanced activity illustrated as a mat activity is that of "Bipedal Walking" (Fig. 75). In general, prone progression, or locomotion, is best performed on a mat. Bipedal activities in the upright position which require a flat surface also may be performed on a mat. In fact, in walking barefoot on a mat, response of the feet is not hampered by close-fitting shoes and action of the feet may be encouraged and observed more easily.

When the total structure is involved, the total patterns of developmental activities are used to hasten learning of total patterns of movement. When one segment is involved and the remainder of the total structure is intact, total patterns provide optimum reinforcements for the deficient segment. Any contraindications for weight-bearing must be heeded. Given an intact skeletal system and integrity of joint structures, the use of total patterns promotes increasing ranges of motion and establishment of the proper interaction of antagonistic component patterns and interaction of segments.

Fig. 38. Head and neck: flexion with rotation.

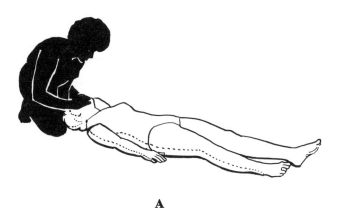

A

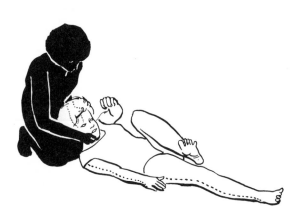

B

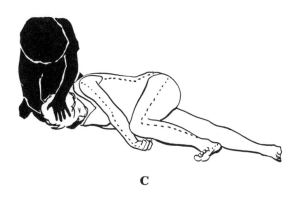

C

Component patterns

RESISTED

Head and neck, flexion with rotation to right

FREE

Left upper extremity, extension–adduction–internal rotation

Left lower extremity, flexion–adduction–external rotation

Right extremities adjust toward extension and adduction

A. Lengthened range

COMMANDS

"Lift your head, look at your right hip, and roll!"
"Pull your left arm down and across!"
"Pull your left foot up and across!" "Roll!"

SUGGESTED TECHNIQUES

Traction, stretch; and resistance

B. Approaching middle range

COMMANDS

"Pull your chin down some more!"
"Roll on over!"

SUGGESTED TECHNIQUES

Repeated contractions, or if patient cannot complete roll, slow reversal followed by repeated contractions

C. Approaching shortened range

COMMANDS

"Hold it! Don't let me pull you back!"

SUGGESTED TECHNIQUES

Repeated contractions, rhythmic stabilization (see Balance—Sidelying, Fig. 44), slow reversal, slow reversal-hold

Antagonistic Pattern

Rolling from prone toward supine—Head and neck extension with rotation to left (Fig. 45)

Left extremities move in antagonistic patterns (see Fig. 45, *Note*)

Note: Intermediate joints, elbow and knee, of moving extremities may flex, extend, or remain straight.

Fig. 39. Head and neck: flexion with rotation, contralateral scapula.

Component patterns

RESISTED

Head and neck, flexion with rotation to right
Left upper extremity, extension–adduction–internal rotation, scapula depresses anteriorly

FREE

Left lower extremity, flexion–adduction–external rotation
Right extremities adjust in extension and adduction

A. Lengthened range

COMMANDS

"Pull your chin to your chest, pull your arm down and across, and roll over!"
"Pull your left foot up and across!" "Roll!"

SUGGESTED TECHNIQUES

Stretch and resistance

B. Approaching middle range

COMMANDS

"Reach for your right hip!" "All the way!"

SUGGESTED TECHNIQUES

Repeated stretch, repeated contractions, or slow reversal followed by renewed effort to roll

C. Approaching shortened range

COMMANDS

"Hold it!" "Pull again!"

SUGGESTED TECHNIQUES

Repeated contractions, rhythmic stabilization (see Balance—Sidelying, Fig. 44), slow reversal, slow reversal—hold

Antagonistic Pattern

Rolling from prone toward supine—Head and neck extension with rotation to left; left upper extremity flexion–abduction–external rotation, scapula elevates posteriorly
Left lower extremity moves in antagonistic pattern

Note: Intermediate joints, elbow and knee, of moving extremities may flex, extend, or remain straight.

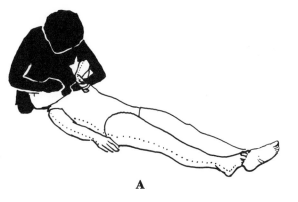

A

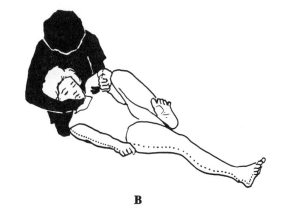

B

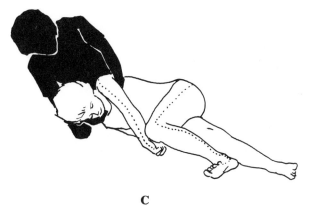

C

Fig. 40. Head and neck: flexion with rotation, bilateral asymmetrical upper extremities.

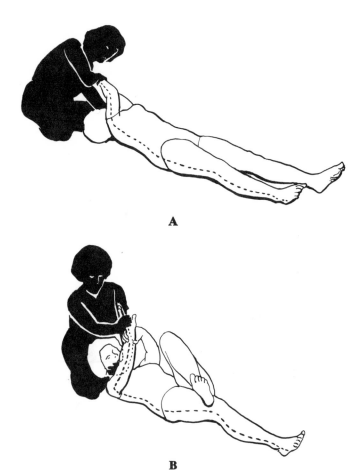

A

B

C

Component patterns

RESISTED

Head and neck, flexion with rotation to right
Upper extremities, asymmetrical extension (chopping) to right

FREE

Left lower extremity, flexion–adduction–external rotation
Right lower extremity adjusts in extension and adduction

A. Lengthened range

COMMANDS

"Pull your arms down toward your right hip, lift your head, and roll over!"
"Pull your left foot up and over!" "Roll!"

SUGGESTED TECHNIQUES

Traction to upper extremities, stretch and resistance

B. Approaching middle range

COMMANDS

"Pull your arms down!" "Roll!"
"Pull your knee on over!" "Roll!"

SUGGESTED TECHNIQUES

Repeated stretch, repeated contractions

C. Approaching shortened range

COMMANDS

"Hold! Don't let me pull you back!"

SUGGESTED TECHNIQUES

Slow reversal followed by repeated contractions, rhythmic stabilization

Antagonistic Pattern

Rolling from prone toward supine—Head and neck extension with rotation to left; bilateral asymmetrical flexion of upper extremities (lifting) to left (Fig. 46)
Left lower extremity moves in antagonistic pattern

Note: Intermediate joints, elbows and knee, of moving extremities may flex, extend, or remain straight.

MAT ACTIVITIES
Rolling: Supine toward Prone

Fig. 41. Head and neck: rotation.

Component patterns

RESISTED

Head and neck, rotation to right
Lower trunk, rotation to right

FREE

Left upper extremity, flexion–adduction–external rotation
Left lower extremity, flexion–adduction–external rotation
Right extremities adjust in extension and adduction.

A. Lengthened range

COMMANDS

"Turn your head toward your right shoulder, pull your hand across your face, and roll!"
"Lift your left foot up and across!" "Roll over!"

SUGGESTED TECHNIQUES

Stretch and resistance

B. Approaching middle range

COMMANDS

"Open your hand now and reach for the mat!"
"Pull your left foot across! Roll!"

SUGGESTED TECHNIQUES

Repeated stretch, repeated contractions

C. Approaching shortened range

COMMANDS

"Pull your left hip down to the mat!"
"Hold it! Don't let me pull you back!"

SUGGESTED TECHNIQUES

Repeated contractions, rhythmic stabilization (see Balance—Sidelying, Fig. 44)

Antagonistic Pattern

Rolling from prone toward supine—Head and neck rotation to left; lower trunk rotation to left
Lower extremities move in antagonistic patterns (see Figs. 47 and 48)

Note: Head and neck rotation, as with all patterns of head and neck, activates the related trunk patterns. In this total pattern of rolling, upper and lower trunk

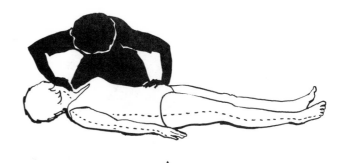

A

B

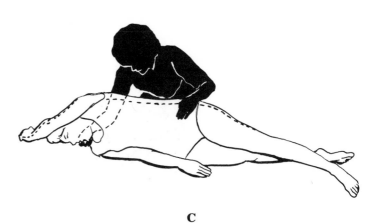

C

rotate from lateralward extension on the left to lateralward extension on the right, as do head and neck.

Left upper extremity may thrust with opening of hand toward ulnar side and with elbow, initially positioned in flexion, extending. Knee may flex or extend.

Fig. 42. Head and neck: extension with rotation, contralateral lower extremity flexion.

A

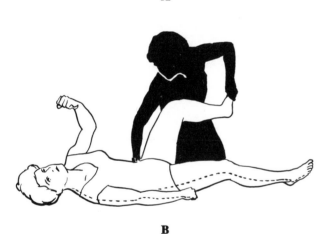

B

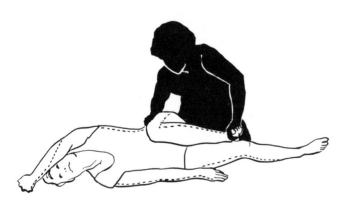

C

Component patterns

RESISTED

Left lower extremity, flexion–adduction–external rotation.

FREE

Head and neck, extension with rotation to right
Left upper extremity, flexion–adduction–external rotation
Right extremities adjust in extension and adduction

A. Lengthened range

COMMANDS

"Look up over your right shoulder, pull your foot up and bend your knee, and roll!"
"Reach your hand up and across your face!" "Roll!"

SUGGESTED TECHNIQUES

Traction, stretch and resistance

B. Approaching middle range

COMMANDS

"Bend your knee and pull it across!"
"Reach for the mat, turn your head!" "Roll!"

SUGGESTED TECHNIQUES

Slow reversal followed by renewed effort to roll

C. Approaching shortened range

COMMANDS

"Pull your knee down to the mat, and hold it!"

SUGGESTED TECHNIQUES

Repeated contractions, slow reversal, slow reversal—hold

Antagonistic Pattern

Rolling from prone toward supine—Left lower extremity, extension–abduction–internal rotation; head and neck flexion with rotation to left

Left upper extremity moves in antagonistic pattern (see *Note*, Fig. 43)

Note: In *A*, subject may be asked to lift head and look at left foot before extending head and neck with rotation to right. Left upper extremity may thrust in flexion–adduction–external rotation (see *Note*, Fig. 41). Note that flexion–adduction–external rotation patterns of left extremities have been combined with head and neck extension with rotation, and with head and neck rotation patterns (see Fig. 41).

Fig. 43. Head and neck: extension with rotation, contralateral lower extremity extension.

Component patterns

RESISTED

Left lower extremity, extension–abduction–internal rotation

FREE

Head and neck, extension with rotation to right
Left upper extremity, flexion–adduction–external rotation
Right extremities adjust in extension and adduction.

A. Lengthened range

COMMANDS

"Look up to your right, push your foot to me, and roll over!"
"Turn your head, look up, and push!"

SUGGESTED TECHNIQUES

Stretch and resistance

B. Approaching middle range

COMMANDS

"Push to me! Roll over!"

SUGGESTED TECHNIQUES

Repeated stretch

C. Approaching shortened range

COMMANDS

"Straighten your knee!"

SUGGESTED TECHNIQUES

See *Note*.

Antagonistic Pattern

See *Note*.

Note: An adversive movement, extension of the lower extremity in a thrusting pattern, has been used to initiate rolling toward prone. The normal child may initiate rolling by pushing against the surface. The subject, in this instance, pushes against the therapist's hand. The subject, as with the child, could push against the surface (mat) while the therapist resisted other component patterns.

The extension–abduction–internal rotation pattern may be used to initiate rolling from prone toward supine. This may be visualized by superimposing the position of the left lower extremity, as shown in *A,* on the position of head and trunk, as shown in *C*. In Figures 45–48, the extension–abduction–internal rota-

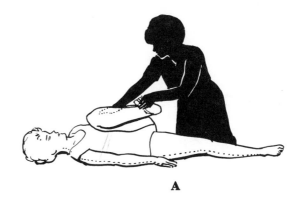

A

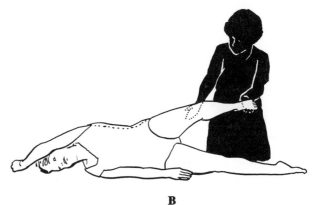

B

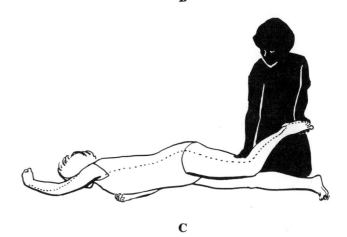

C

tion pattern of the lower extremity is used consistently in rolling from prone to supine, while in Figures 38–42 the flexion–adduction–external rotation pattern of the lower extremity is used consistently in rolling from supine toward prone. The opposite diagonal of lower extremity patterns is not used to promote rolling.

Fig. 44. Sidelying balance.

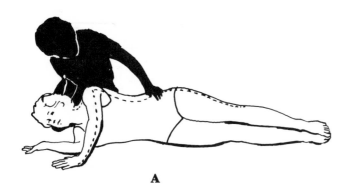

A

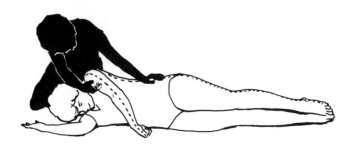

B

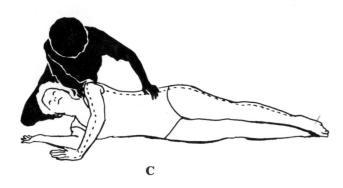

C

Component patterns

A. RESISTED

Head and neck, extension with rotation to right
Lower trunk, rotation

FREE

Left upper extremity stabilizes in diagonal of flexion–adduction–external rotation and extension–abduction–internal rotation patterns
Right upper extremity stabilizes against surface as necessary
Lower extremities in close approximation

B. RESISTED

Left upper extremity, extension–adduction–internal rotation, scapula depressed anteriorly
Lower trunk, rotation

FREE

Head and neck, flexion with rotation to right
Lower extremities in close approximation

C. RESISTED

Left upper extremity, flexion–adduction–external rotion, scapula elevated anteriorly
Lower trunk, rotation

COMMANDS

A and *B,* "Hold, don't let me move you in any direction."
C, "Hold. Don't let me pull you back!"

SUGGESTED TECHNIQUES

A and *B,* rhythmic stabilization of flexion and extension components at the same time
C, rhythmic stabilization of flexion, then extension components

Note: Balancing, maintaining a position or posture, requires stability. Balancing against resistance promotes stability especially when antagonistic patterns are resisted at the same time, as in *A* and *B*. As performed in *C*, stability is threatened so that balance must be recovered by isotonic contraction in compensatory movements.

The sidelying or lateral position may be used to initiate rolling toward supine or toward prone where the patient cannot initiate against resistance from supine or prone positions. Because gravity will assist in either direction, weak patterns may be resisted.

Fig. 45. Head and neck: extension with rotation.

Component patterns

RESISTED

Head and neck, extension with rotation to left

FREE

Left upper extremity, extension–abduction–internal rotation

Left lower extremity, extension–abduction–internal rotation

Right extremities adjust in extension and abduction

A. Lengthened range

COMMANDS

"Look up at me, lift your head back to me, and roll!"
"Push with your left hand!" "Lift your leg back! Roll!"

SUGGESTED TECHNIQUES

Stretch and resistance

B. Middle range

COMMANDS

"Push down to the mat!" "Look back here!"

SUGGESTED TECHNIQUES

Slow reversal, repeated contractions

C. Shortened range

COMMANDS

"Hold it! Don't let me lift you from the mat!"

SUGGESTED TECHNIQUES

Slow reversal

Antagonistic Pattern

Rolling from supine toward prone—Head and neck, flexion with rotation

Right upper extremity moves in extension–adduction–internal rotation pattern (Fig. 38)

Left lower extremity moves in antagonistic pattern.

Note: Intermediate joints, elbow and knee, of moving extremities extend or remain straight.

In this instance, because of the position of the therapist, the subject has used the left upper extremity in the extension–abduction–internal rotation pattern rather than the more closely related flexion–adduction–external rotation pattern as combined for "lifting" as in

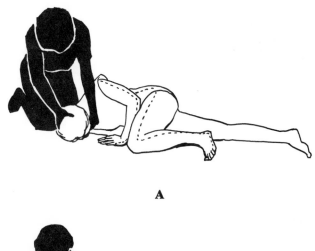

A

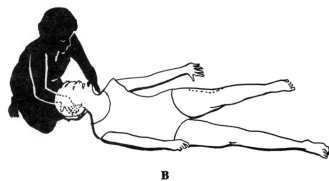

B

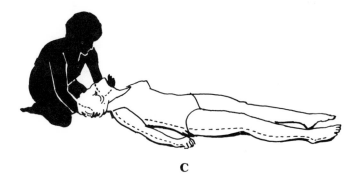

C

Figure 46. Therefore, the combinations used in Figures 38 and 45 are not directly antagonistic for the left upper extremity.

Use of the developmental sequence

Fig. 46. Head and neck: extension with rotation, bilateral asymmetrical upper extremities.

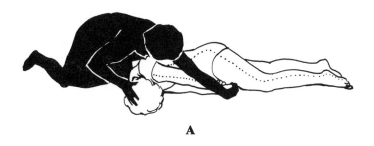

A

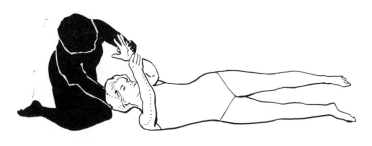

B

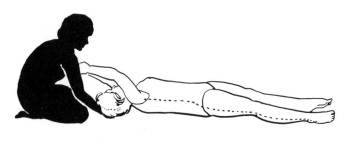

C

Component patterns

RESISTED

Head and neck, extension with rotation to left
Upper extremities, asymmetrical flexion (lifting) to the left

FREE

Left lower extremity, extension–abduction–internal rotation
Right lower extremity adjusts in extension and adduction.

A. Lengthened range

COMMANDS

"Look at me, lift your head and hands to me, and roll!"
"Lift your left leg back!" "Roll!"

SUGGESTED TECHNIQUES

Traction to upper extremities, stretch and resistance

B. Middle range

COMMANDS

"Look back at me!" "Lift your arms!"

SUGGESTED TECHNIQUES

Rhythmic stabilization, repeated contractions, slow reversal, slow reversal–hold

C. Shortened range

COMMANDS

"Hold! Don't let me lift you!"

SUGGESTED TECHNIQUES

Approximation to upper extremities, repeated contractions

Antagonistic Pattern

Rolling from supine toward prone—Head and neck flexion with rotation to right; bilateral asymmetrical extension of upper extremities (chopping) to right (Fig. 40)
Left lower extremity moves in antagonistic pattern.

Note: Intermediate joints of upper extremities may flex, extend, or remain straight. Approximation is used with extending or extended elbows.

If the position of head and upper extremities, as shown in *A,* is superimposed upon the supine body position in *C,* the use of this combination to initiate rolling from supine toward prone to the left can be visualized. Thus, extension with rotation of head and neck may be used to promote rolling toward supine, and rolling toward prone.

Rolling: Prone toward Supine

Fig. 47. Head and neck: rotation, ipsilateral upper extremity.

Component patterns

RESISTED

Head and neck, rotation to the left
Left upper extremity, extension–abduction–internal rotation

FREE

Left lower extremity, extension–abduction–internal rotation
Right extremities adjust in extension and abduction

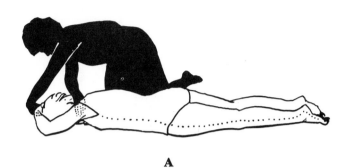

A

A. Lengthened range

COMMANDS

"Open your hand, turn your head, and roll toward me!"
"Lift your left leg back to the mat!" "Roll!"

SUGGESTED TECHNIQUES

Stretch and resistance

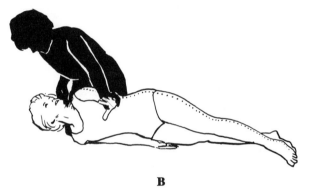

B

B. Approaching middle range

COMMANDS

"Hold! Now, push on back here!"
"Lift that foot back here!"

SUGGESTED TECHNIQUES

Repeated contractions, rhythmic stabilization (see Balance—Sidelying, Fig. 44)

C. Approaching shortened range

COMMANDS

"Push some more! Hold it! Push again!"

SUGGESTED TECHNIQUES

Approximation to left upper extremity followed by repeated contractions, slow reversal, slow reversal—hold

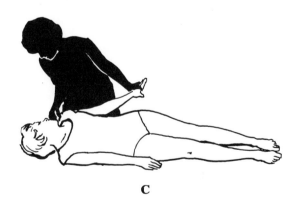

C

Antagonistic pattern

Rolling from supine toward prone—Head and neck rotation to right
Left extremities move in antagonistic patterns (Fig. 41)

Note: Intermediate joints, elbow and knee, of moving extremities extend or remain straight. If, as the middle range is approached, *B*, the subject were to flex the left knee, the foot could contact the mat so as to thrust in the extension–abduction–internal rotation pattern as described in *Note,* Figure 43.

Fig. 48. Head and neck: rotation, ipsilateral scapula and pelvis.

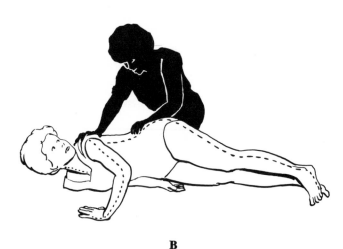

A

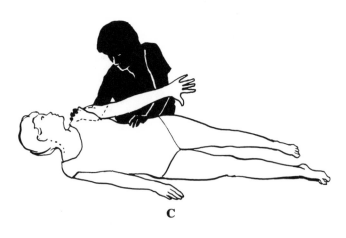

B

C

Component patterns

RESISTED

Left upper extremity, extension–abduction–internal rotation, scapula depresses posteriorly
Lower trunk rotation to left

FREE

Head and neck, rotation
Left lower extremity, extension–abduction–internal rotation
Right extremities adjust in extension and abduction

A. Lengthened range

COMMANDS

"Turn your head to me, push with your left hand, and roll!"
"Lift your left leg back toward me!" "Roll!"

SUGGESTED TECHNIQUES

Stretch and resistance

B. Approaching middle range

COMMANDS

"Push!" "Roll back here!"

SUGGESTED TECHNIQUES

Approximation to left upper extremity, repeated contractions, slow reversal-hold

C. Approaching shortened range

COMMANDS

"Come on back! Turn all the way!"

SUGGESTED TECHNIQUES

Rhythmic stabilization (see Balance—Sidelying, Fig. 44), repeated contractions, slow reversal, slow reversal-hold

Antagonistic pattern

Rolling from supine toward prone—Head and neck rotation to right; lower trunk rotation to right
Left extremities move in antagonistic patterns (Fig. 41)

Note: Intermediate joints, elbow and knee, of moving extremities extend, or remain straight.
In *A,* subject could have placed left hand near her forehead so as to more specifically activate the extension–abduction–internal rotation pattern of the left upper extremity.

MAT ACTIVITIES
Activities for Lower Trunk (Inferior Region)

Fig. 49. Lower trunk: rotation, supine.

Component patterns

RESISTED

Lower trunk, rotation (with hips and knees flexed) to right

FREE

Head and neck, rotation
Upper extremities adjust in extension and abduction

A. Lengthened range

COMMANDS

"Turn your head and knees to the right!"
"Pull away from me!"

SUGGESTED TECHNIQUES
Stretch and resistance

B. Middle range

COMMANDS

"Turn your head, and pull your knees on over!"

SUGGESTED TECHNIQUES

Rhythmic stabilization followed by repeated contractions

C. Shortened range

COMMANDS

"Hold! Don't let me pull you back!"

SUGGESTED TECHNIQUES

Repeated contractions, slow reversal, slow reversal-hold

Antagonistic pattern

Lower trunk rotation to left

Note: If necessary, lower trunk rotation with bilateral asymmetrical flexion of lower extremities may be used to initiate the ability to roll from supine toward prone. For completion of rolling, the left upper extremity should contribute with the flexion–adduction–external rotation pattern (see Figure 41). If the head is rotated to the left while lower trunk is rotated to the right, adversive movements produce stability of trunk.

As with head and neck rotation, lower trunk rotation proceeds from lateralward extension on the left, through flexion, to lateralward extension on the right. In *B*, subject has flexed head and neck as flexion phase is active in lower trunk. Usually, head and neck rotate without lifting from contact with surface.

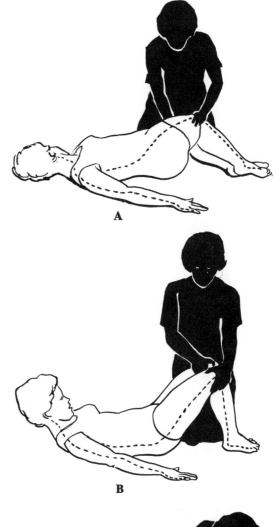

A

B

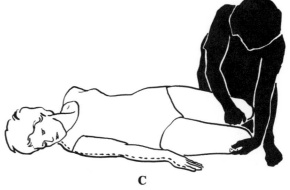

C

Lower trunk rotation, supine, activates extensors. If performed in prone position (hips extended, knees flexed), flexors are activated.

MAT ACTIVITIES
Activities for Lower Trunk (Inferior Region)

Fig. 50. Pelvic elevation, supine.

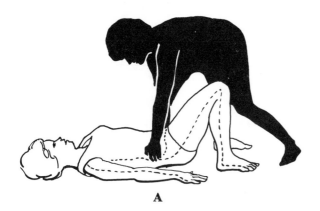

A

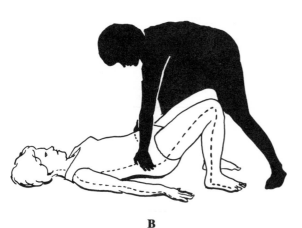

B

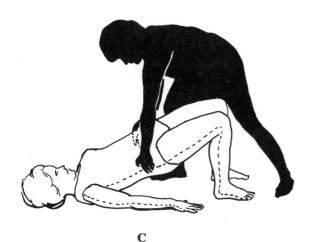

C

Component patterns

RESISTED

Pelvic elevation with extension of lower trunk (Bridging)

FREE

Head and neck adjust in mid-position
Upper extremities adjust in extension and abduction

A. Lengthened range

COMMANDS

"Push with your head and your feet, and lift your hips!"

SUGGESTED TECHNIQUES

Stretch and resistance

B. Middle range

COMMANDS

"Keep on pushing up! Hold it! Push again!"

SUGGESTED TECHNIQUES

Repeated contractions, rhythmic stabilization, slow reversal

C. Shortened range

COMMANDS

"Hold it! Don't let me push you down!"

SUGGESTED TECHNIQUES

Rhythmic stabilization, slow reversal

Antagonistic Pattern

Reversal to supine position with hips and knees flexed

Note: This activity is resisted diagonally by commanding the subject to push toward one side while resistance is graded to promote range toward that side.

Pelvic elevation supine, if viewed by rotating illustration 90 degrees, can be seen to be related to kneeling position (see Figure 69). Total pattern would be that of rising from heel-sitting, feet plantar flexed, to kneeling.

Facilitation of total patterns

MAT ACTIVITIES
Prone Progression: Crawling

Fig. 51. Crawling forward on elbows.

Component patterns

RESISTED

Lower extremities, flexion–abduction–internal rotation, alternately

FREE

Head and neck, flexion with rotation toward left as left lower extremity flexes with abduction, toward right as right lower extremity advances

Upper extremities adjust alternately between flexion with abduction and extension with adduction

Lower extremity which has flexed then adjusts in extension–adduction–external rotation as opposite extremity flexes

A. Lengthened range—Left

COMMANDS

"Pull your left foot and your knee up and out, and pull yourself forward!"

SUGGESTED TECHNIQUES

Traction, stretch, and resistance

B. Approaching middle range—Left

COMMANDS

"Pull your knee forward!"

SUGGESTED TECHNIQUES

Repeated contractions

C. Approaching shortened range—Right

COMMANDS

"Pull yourself forward! Use your hands!"

SUGGESTED TECHNIQUES

Slow reversals of right and left extremities alternately

Antagonistic pattern

Crawling backward on elbows (Fig. 52)

Note: The amount of resistance used will influence movement of head and neck. With less resistance subject would be inclined to maintain head and neck in extension.

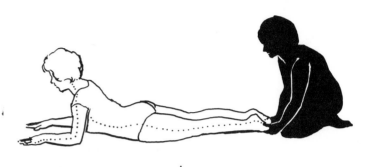

A

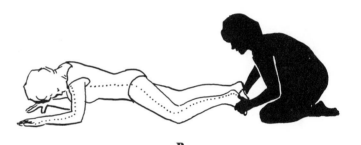

B

C

In the most primitive form, head and neck are not elevated so that rotation from one side to the other is used. Head turns to left when left upper and lower extremities advance in flexion with abduction, an ipsilateral combination. Other combining movements include: alternating reciprocal of upper, then lower extremities; and diagonal reciprocal with contralateral upper and lower extremities advancing at the same time.

Fig. 52. Crawling backward on elbows.

FREE

Head and neck, extension with rotation toward left as left lower extremity extends with adduction, toward right as right lower extremity extends with adduction

Upper extremities adjust alternately between extension with adduction and flexion with abduction

Lower extremity which has extended then adjusts in flexion after opposite extremity extends

A. Middle range—Left

COMMANDS

"Push with your hands, and push your left foot back to me!"

SUGGESTED TECHNIQUES

Stretch and resistance

B. Lengthened range—Right

COMMANDS

"Now! Push your right foot back, and push with your hands!"

SUGGESTED TECHNIQUES

Stretch and resistance

C. Approaching shortened range—Right

COMMANDS

"Push all the way!"

SUGGESTED TECHNIQUES

Slow reversals of left and right extremities alternately

Antagonistic pattern

Crawling forward on elbows (Fig. 51)

Note: See *Note,* Figure 51.

Crawling forward and backward on elbows requires elevation of superior region, head and neck and thorax. Elevation of superior region is assumed, as shown in Figure 53, *A*. Rhythmic stabilization may be used to promote head control and stability of shoulder girdle.

When a diagonal direction is attempted, the total pattern resembles circular pivoting (prone). Circular pivoting should be performed first without elevation of superior region so that total body is in contact with surface.

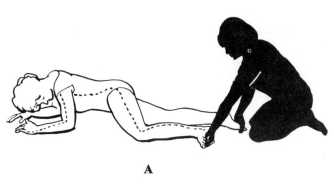

A

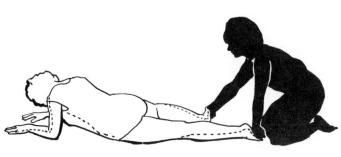

B

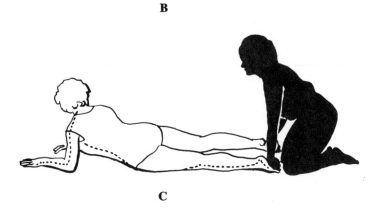

C

Component patterns

RESISTED

Lower extremities, extension–adduction–external rotation, alternately

Facilitation of total patterns

Prone Progression: On Elbows and Knees

Fig. 53. Rising to elbows and knees.

Component patterns

RESISTED

Head and neck, extension with rotation to right
Lower trunk, flexion with rotation toward right

FREE

Upper extremities adjust with bilateral asymmetrical flexion toward left
Lower extremities adjust with flexion toward right

A. Lengthened range

COMMANDS

"Lift up to me!" "Get your head up! And, your hips!"

SUGGESTED TECHNIQUES

Stretch and resistance

B. Middle range

COMMANDS

"Hold it! Now, push some more!"

SUGGESTED TECHNIQUES

Repeated contractions

C. Approaching shortened range

COMMANDS

"Move your left arm back a little. Now, your right."
"Now, push your hips back!"

SUGGESTED TECHNIQUES

Resistance

Antagonistic pattern

Reversal to prone position

Note: Diagonal direction promotes elbow extension on the left when total movement is backward to the right. Rocking forward and backward, and diagonally, develops ability to maintain balance with trunk free of contact with surface.

If necessary, as an intermediate activity, the superior region may remain in contact with surface while the inferior region, lower trunk, is resisted, or assisted to a chest and knees position. Rocking movements of the inferior region in all directions, and rhythmic stabilization may be used to promote stability of inferior region.

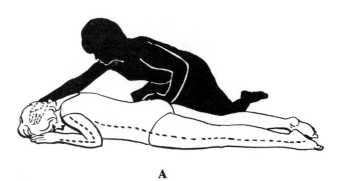

A

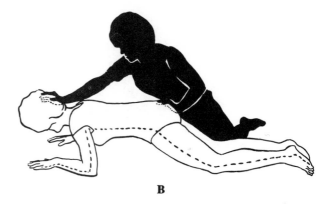

B

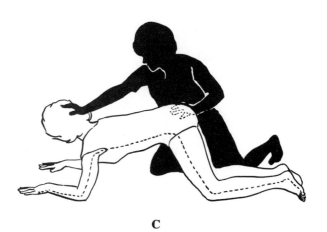

C

Use of the developmental sequence

Fig. 54. Balance on elbows and knees.

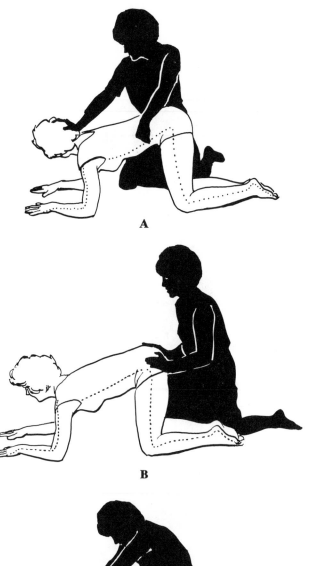

Component patterns

A. RESISTED

Head and neck, extension with rotation to right
Lower trunk, rotation to left

FREE

Upper extremities adjust with asymmetrical flexion toward left
Lower extremities adjust with asymmetrical flexion toward left

B. RESISTED

Lower trunk, flexion with rotation to right

FREE

Head and neck, extension with rotation to right
Upper extremities adjust with asymmetrical flexion toward left
Lower extremities adjust with asymmetrical flexion toward left

C. RESISTED

Lower trunk, extension with rotation toward left

FREE

Head and neck, flexion with rotation toward left
Upper extremities adjust with asymmetrical extension toward right
Lower extremities adjust with asymmetrical extension toward left

COMMANDS

A, "Hold! Don't let me push your head down! Don't let me pull your hips toward me! Stay there!"
B, "Don't let me push you forward! Hold still!"
C, "Don't let me pull you back! Hold!"

SUGGESTED TECHNIQUES

A, rhythmic stabilization of upper and lower trunk
B, rocking forward toward extension, and backward toward flexion, then rhythmic stabilization
C, holding alternately with hip extensors, then flexors

Note: In *A,* rotation components combined with extension of upper trunk and with flexion of lower trunk are resisted at the same time so that stability is achieved. In *B,* flexion of lower trunk is necessary to maintain position. If pelvis is pulled toward heels by therapist, extension of lower trunk may be resisted. In *C,* extension of lower trunk is necessary to maintain position. If therapist pushes subject forward from scapular and axillary regions, subject will elevate head and will flex lower trunk. In *B* and *C,* stability is threatened alternately; disturbance of equilibruim promotes compensatory movements with isotonic contractions of muscle groups in order to recover position. See Figure 57 for other combinations of manual contacts.

MAT ACTIVITIES
Prone Progression: On Hands and Knees

Fig. 55. Rising to hands and knees.

Component patterns

RESISTED

Head and neck extension with rotation to right
Left upper extremity, flexion–abduction–external rotation

FREE

Right upper extremity, flexion–adduction–external rotation
Lower trunk, flexion with rotation to left
Lower extremities adjust with flexion toward left

A. Lengthened range

COMMANDS

"Lift your head toward me, push with your hands, and push back to me!"

SUGGESTED TECHNIQUES

Stretch and resistance

B. Middle range

COMMANDS

"Come on up, on your hands and knees!"

SUGGESTED TECHNIQUES

Repeated contractions, slow reversal

C. Shortened range

COMMANDS

"Hold!" Straighten your elbows!"

SUGGESTED TECHNIQUES

Approximattion to left upper extremity, rhythmic stabilization

Antagonistic pattern

Reversal to prone position

Note: Emphasis is on extension of elbows. Subject could assume elbows and knees position first (see Fig, 53, *C*), and then rise to hands and knees by extending one elbow first, and then the other.

For complete verticality of thighs and upper extremities, subject must adjust upper extremities toward extension. Rocking toward one side and then the other permits the adjustment to be made. See Figures 54, 56, and 57 for related balancing activities and a variety of manual contacts.

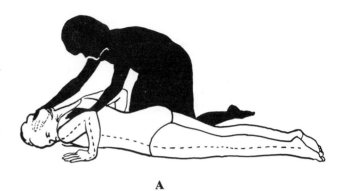

A

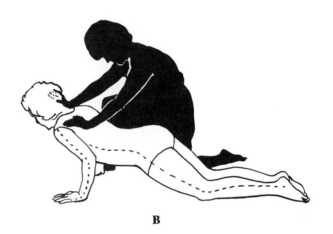

B

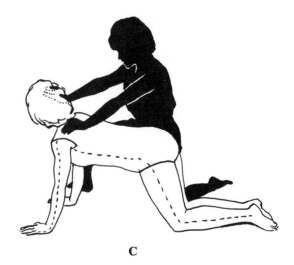

C

Fig. 56. Rocking on hands and knees.

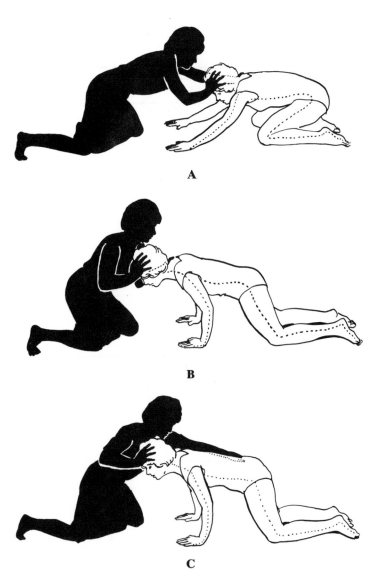

A

B

C

Component patterns

RESISTED

Head and neck, extension with rotation toward left

FREE

Lower trunk, extension with rotation toward right
Upper extremities adjust with asymmetrical extension toward right
Lower extremities adjust with extension toward right

A. Lengthened range

COMMANDS

"Push yourself forward toward me!"

SUGGESTED TECHNIQUES

Approximation of trunk, stretch and resistance

B. Approaching shortened range

COMMANDS

"Hold! I'm going to push you back!"

SUGGESTED TECHNIQUES

Slow reversal, slow reversal—hold

C. Approaching middle range

COMMANDS

"Straighten your elbows, and hold!"

SUGGESTED TECHNIQUES

Rhythmic stabilization, slow reversal (forward and backward)

Antagonistic Pattern

Rocking forward is antagonistic to rocking backward and the reverse

Note: In *A*, head is positioned in middle range rather than in lenghthened range of extension so as to use approximation and stretch of hip extensors.

Rocking may be performed through extreme ranges, from position in *A* forward and beyond position in *B;* or, rocking may be performed through brief excursions of range. Performance with decrements in range using slow reversal—hold technique promotes balance at middle range, the hands and knees position.

The position shown in *A* may be used for forward assumption of hands and knees position by extension of inferior region, while in Figure 55 position was assumed by flexion of inferior region and moving backward.

Acts of assumption, such as rolling into the side-lying position, rising to sitting, rising to hands and knees enhance the ability to maintain the assumed position. Rocking movements, performed in all possible directions, further enhance the ability to locate the point of balance and to recover equilibrium when it is disturbed deliberately or inadvertently.

Prone Progression: On Hands and Knees

Fig. 57. Balance on hands and knees.

Component patterns

A. RESISTED

Head and neck, flexion with rotation to left
Lower trunk, extension with rotation toward right

FREE

Upper extremities adjust with asymmetrical extension toward left

B. RESISTED

Left upper extremity, flexion–abduction–external rotation
Lower trunk, rotation to right

FREE

Head and neck adjust in extension toward left
Right upper extremity adjusts with flexion–adduction–external rotation
Lower extremities adjust with asymmetrical flexion toward right

C. RESISTED

Left upper extremity, flexion–abduction–external rotation
Lower trunk, flexion with rotation to right

FREE

Head and neck, flexion with rotation to left
Right upper extremity, flexion–adduction–external rotation
Lower extremities adjust with flexion toward right

COMMANDS

A, "Hold! Don't let me left you up! Nor, push you back!"
B, "Hold, and keep your left hand and knee on the mat."
C, "Don't let me push your hips to the left!"

SUGGESTED TECHNIQUES

A, rhythmic stabilization
B, stretch with disturbance so as to demand that subject reach for mat
C, approximation of trunk so as to promote elevation of head

A

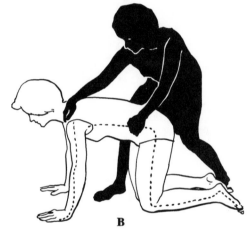

B

C

Note: In *A* and *B,* subject may be disturbed sufficiently to promote reaching for mat with open hand and with right knee, *A,* and left knee, *B.* For stability it is necessary to perceive weight bearing on the supporting segments. In *C,* superior region is disturbed from left to right while inferior region is disturbed from right to left. Because counter-pressures and adversive directions are used, subject becomes stable. See Figures 54 and 56 for other appropriate manual contacts.

Fig. 58. Creeping forward to left, ipsilateral scapula and pelvis.

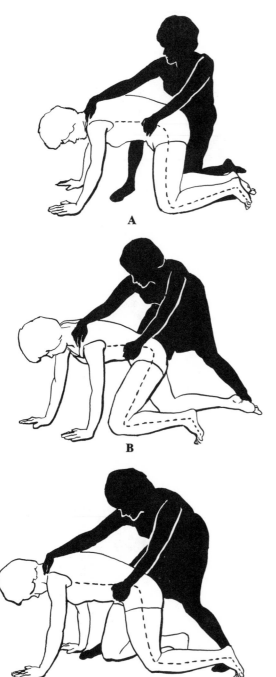

A

B

C

Component patterns

RESISTED

Left upper extremity, flexion–abduction–external rotation

Left lower extremity, flexion–abduction–internal rotation

FREE

Right upper extremity, flexion–adduction–external rotation

Right lower extremity, flexion–adduction–external rotation

Head and neck adjust as necessary

A. Initial position

COMMANDS

"Move your right arm forward to the left."

SUGGESTED TECHNIQUES

Prepare to resist movement forward to left

B. Shortened range—Left lower extremity

COMMANDS

"Pull your left foot and knee forward to the left!"
"Now, move your left arm forward and sideward!"

SUGGESTED TECHNIQUES

Resistance, rhythmic stabilization

C. Shortened range—Right lower extremity

COMMANDS

"Pull your right knee toward your left shoulder!"

SUGGESTED TECHNIQUES

Decrease resistance at pelvis so that subject bears more weight on left knee

Antagonistic pattern

Creeping backward to right (Fig. 61).

Note: Subject fails to elevate head so as to look in direction of movement because therapist is pulling diagonally backward yet is lifting the subject away from the surface so that trunk flexors are activated. Head and neck adjust accordingly. Manual contacts favor flexion–abduction patterns of left upper and lower extremities which must advance if progression is to occur. At the same time, the subject is inclined to use extension of the right hip and knee.

Sequence is that of diagonal reciprocation: right upper, left lower, left upper, right lower extremities. Other less advanced combinations may be used: ipsilateral with left upper then left lower advancing, or with left extremities advancing together; bilateral asymmetrical placement of upper extremities followed by alternating reciprocal of lowers, left then right; or alternating reciprocal of upper extremities followed by alternating reciprocal of lower extremities.

Facilitation of total patterns

Prone Progression: On Hands and Knees

Fig. 59. Creeping forward to left, pelvis.

Component patterns

RESISTED

Left lower extremity, flexion–abduction–internal rotation

Right lower extremity, flexion–adduction–external rotation

FREE

Head and neck adjust in extension

Right upper extremity, flexion–adduction–external rotation

Left upper extremity, flexion–abduction–external rotation

A. Initial position

COMMANDS

"Hold! Now, move your right arm forward to the left."

SUGGESTED TECHNIQUES

Rhythmic stabilization to rotation of pelvis

B. Shortened range—Left lower extremity

COMMANDS

"Put your weight on your left knee, and move your right arm forward to the left."

SUGGESTED TECHNIQUES

Resistance

C. Shortened range—Right lower extremity

COMMANDS

"Pull your right knee toward your left shoulder!"

SUGGESTED TECHNIQUES

Increase resistance on right, pulling upward and backward

Antagonistic pattern

Creeping backward to right (Fig. 61)

Note: Because, as compared with Figure 58, there is less activation of trunk flexors, head and neck adjust with extension.

Other combinations of extremity movements may be used as described in *Note*, Fig. 58. Other appropriate manual contacts include resistance applied at head and neck with head and neck extension resisted during diagonal creeping forward and backward, and with head and neck flexion resisted during creeping backward. Appropriate contacts with head and mandible

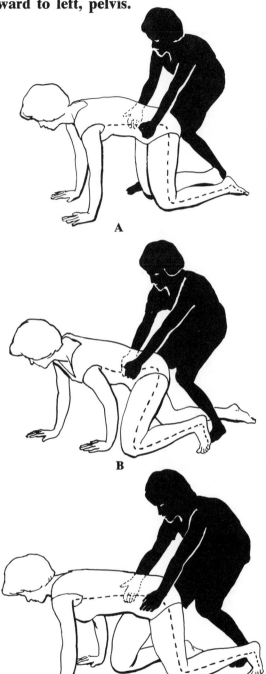

A

B

C

should be used. Manual contacts may be used at shoulders bilaterally, or at head and shoulder as in Figure 56, *A* and *B;* or at head and pelvis, as in *C*.

All manual contacts may be adjusted for creeping diagonally backward, or sideward, or in a circle—head, head and shoulder, shoulders, shoulder and pelvis, and pelvis.

Fig. 60.   Creeping forward to left, lower extremities.

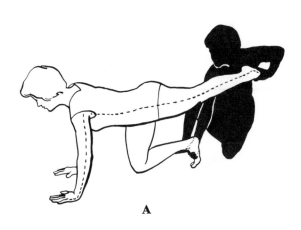

A

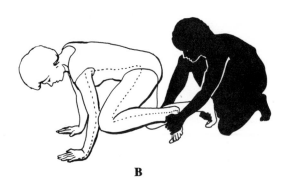

B

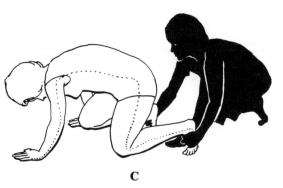

C

Component patterns

RESISTED

Left lower extremity, flexion–abduction–internal rotation

Right lower extremity, flexion–adduction–external rotation

FREE

Head and neck adjust from extension toward flexion
Upper extremities stabilize and adjust to movement of lower extremities

A. Lengthened range—Left lower extremity

COMMANDS

"Pull your left foot and knee forward to your left!"
"Pull! Bend your knee!"

SUGGESTED TECHNIQUES

Traction, stretch, and resistance

B. Shortened range—Left lower extremity

COMMANDS

"Hold it! And, let me have your right leg."

SUGGESTED TECHNIQUES

Stretch and resistance to right lower extremity

C. Shortened range—Lower extremities

COMMANDS

"Hold with your right, and let me have your left leg."

SUGGESTED TECHNIQUES

Repeat flexion patterns of lower extremities

Antagonistic pattern

Creeping backward to right (Fig. 61).

Note: Subject has contacted surface with feet only. Contacting surface with knees would be less difficult but would permit less range of flexion patterns.

In this position, lower extremity patterns as used in creeping may be emphasized, although subject does not actually progress forward. Extension patterns may be emphasized in a similar fashion by adjusting manual contacts. Reversal techniques may be used. Also, for greater control of an extremity, therapist may use both hands on one extremity while the other remains stable with knee in contact with surface. See Figure 61 for manual contacts for extension patterns.

Prone Progression: On Hands and Knees

Fig. 61. Creeping backward to right.

Component patterns

RESISTED

Right lower extremity, extension–abduction–internal rotation

Left lower extremity, extension–adduction–external rotation

FREE

Head and neck adjust from flexion toward extension

Left upper extremity, extension–adduction–internal rotation

Right upper extremity, extension–abduction–internal rotation

A. Lengthened range—Right lower extremity

COMMANDS

"Push your right foot back to me!" "Move your left arm back!"

SUGGESTED TECHNIQUES

Stretch and resistance

B. Approaching middle range—Left lower extremity

COMMANDS

"Now, push your left foot back, and then your right arm."

SUGGESTED TECHNIQUES

Resistance

C. Shortened range—Left lower extremity

COMMANDS

"Put more weight on your left knee. Move your left arm back to the right, and push your right foot back toward me."

SUGGESTED TECHNIQUES

Prepare to resist right lower extremity

Antagonistic pattern

Creeping forward to left (Figs. 58–60)

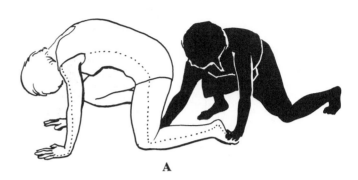

A

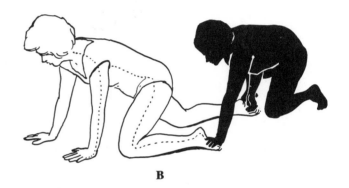

B

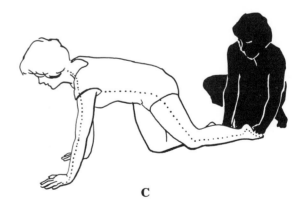

C

Note: Sequence is that of diagonal reciprocation. Other combinations of extremity patterns may be used as described for Figure 58. Extension patterns may be emphasized as for flexion patterns, Figure 60.

For complete development of the total pattern, creeping should be performed in all directions, forward and backward diagonally toward left and right, side-ward to left and right, and in a circle toward left and right. Creeping promotes mass flexion and extension of lower extremities. Plantigrade walking promotes more advanced patterns wherein hip flexion is combined with knee extension and hip extension with knee flexion, Fig. 63).

Use of the developmental sequence   **147**

Prone Progression: On Hands and Feet

Fig. 62. Rising to plantigrade.

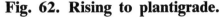

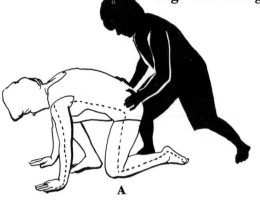

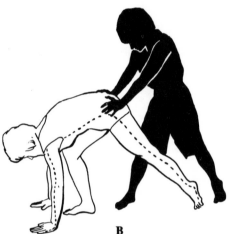

FREE

Head and neck adjust in extension
Upper extremities stabilize and adjust as necessary
Left lower extremity, flexion–adduction–external rotation

A. Initial position—Quadrupedal half-kneeling

COMMANDS

"Push with your right foot, and straighten your knee!"

SUGGESTED TECHNIQUES

Stretch and resistance

B. Middle range

COMMANDS

"Push with your right foot, and pull your left leg forward!"

SUGGESTED TECHNIQUES

Repeated contractions, slow reversal

C. Shortened range

COMMANDS

"Hold it!" "Now, push some more! Straighten your knees!"

SUGGESTED TECHNIQUES

Approximation, rhythmic stabilization to rotation of pelvis

Antagonistic pattern

Reversal to quadupedal half-kneeling

Note: Subject has achieved almost complete extension of knees. Many normal subjects will not extend knees if hips are flexed as is right lower extremity in *B*. Or, if knees are extended, will not flex hips so that both lower extremities assume position of left lower extremity in *B*. In order for plantar surfaces to contact mat, subject should be permitted to flex hips and knees. Then, approximation may be used to encourage further extension of knees. See walking posture, Figure 63.

Plantigrade position may be assumed with bilateral extension of lower extremities by assuming creeping position, as shown in Figure 60, *C*.

Balancing and rocking activities should be carried out.

Squat-sitting and standing position may be assumed from plantigrade posture.

Component patterns

RESISTED

Elevation of pelvis
Right lower extremity, extension–adduction–external rotation

MAT ACTIVITIES
Prone Progression: On Hands and Feet

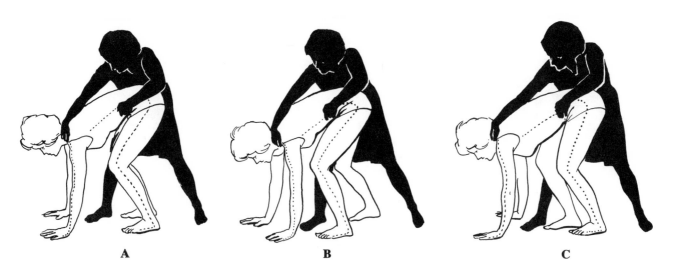

A   B   C

Fig. 63. Plantigrade walking forward to left.

Component patterns

RESISTED

Left upper extremity, flexion–abduction–external rotation

Left lower extremity, flexion–abduction–internal rotation

FREE

Head and neck adjust in extension

Right upper extremity, flexion–adduction–external rotation

Right lower extremity, flexion–adduction–external rotation

A. Initial position

COMMANDS

"Move your right arm forward to the left, and step with your left foot."

SUGGESTED TECHNIQUES

Resistance

B. Shortened range—Left lower extremity

COMMANDS

"Move your left arm forward and outward, push with your right foot, then pull it forward to the left."

SUGGESTED TECHNIQUES

Decreased resistance at pelvis so that subject can bear weight on left

C. Approaching shortened range— Right lower extremity

COMMANDS

"Push with your left foot, and pull your right foot toward your left shoulder."

SUGGESTED TECHNIQUES

Approximation to left lower extremity for increased extension

Antagonistic pattern

Plantigrade walking backward to right

Note: While resistance is applied on the left, the advancement of left extremities is favored. In *B,* as therapist pulls pelvis backward to the right, weight is shifted to right foot so that left lower extremity can advance. Therapist may also use both hands at pelvis, or may move forward and resist superior region, as described in *Note,* Figure 59. For combining movements of extremities, other than diagonal reciprocation, see *Note,* Figure 58. See also *Note,* Figure 61.

Because in crawling and in creeping plantar surfaces of feet do not necessarily contact surface, plantigrade walking is a highly desirable activity preparatory to bipedal walking. Knee extensors begin to cooperate with hip flexors as necessary to swing phase of normal walking.

Fig. 64. Rising to sitting from prone.

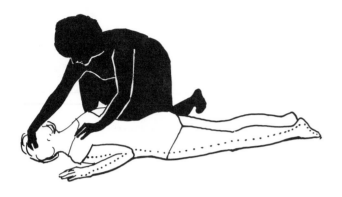

A

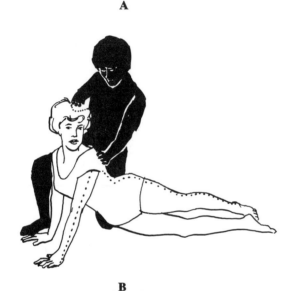

B

C

Component patterns

RESISTED

Head and neck, rotation to left
Upper trunk, rotation to left
Upper extremities, asymmetrical thrusting to right

FREE

Lower trunk, flexion with rotation to right
Lower extremities, asymmetrical flexion to right

A. Lengthened range

COMMANDS

"Turn your head to the left, push with your arms, and sit up toward me!" "Push!"

SUGGESTED TECHNIQUES

Stretch and resistance

B. Middle range

COMMANDS

"Push some more!" Shift your left hand back toward me!" "Turn!" "Hold it!"

SUGGESTED TECHNIQUES

Repeated contractions, slow reversal, approximation to left upper extremity

C. Approaching shortened range

COMMANDS

"Turn all the way, and shift your right hand forward and closer to your hip!"

SUGGESTED TECHNIQUES

Rhythmic stabilization (see Balance—Sitting, Fig. 67), slow reversal

Antagonistic pattern

Reversal to prone position

Note: Assumption of sitting from prone is an asymmetrical total pattern requiring adversive movements of upper and lower trunk. True asymmetry would be evident had subject flexed hips and knees toward the right. Had this occurred, adjustment of lower extremities toward symmetry would have followed symmetrical adjustment of upper extremities and upper trunk.

MAT ACTIVITIES
Sitting

Fig. 65. Rising to sitting from hyperflexion.

Component patterns

RESISTED

Head and neck, extension with rotation to right
Upper trunk, extension with rotation to right
Upper extremities, asymmetrical flexion to right (lifting)

FREE

Lower extremities adjust with bilateral asymmetrical extension of hips

A. Lengthened range

COMMANDS

"Turn your head and lift up away from me! Lift your arms up and away!" "Sit up!"

SUGGESTED TECHNIQUES

Stretch and resistance, traction to upper extremities

B. Middle range

COMMANDS

"Hold it!" "Now, lift up some more! All the way!"

SUGGESTED TECHNIQUES

Repeated contractions, slow reversal

C. Shortened range

COMMANDS

"Hold! Don't let me pull you down!"

SUGGESTED TECHNIQUES

Approximation to upper extremities, rhythmic stabilization (see Balance—Sitting, Fig. 67), slow reversal, slow reversal—hold

Antagonistic pattern

Reversal to hyperflexion

Note: Lower trunk and lower extremities will adjust with asymmetry toward the left in proportion to the extent of rotation occurring in the upper trunk. The knees, which would be inclined toward more flexion in most subjects when trunk is hyperflexed, proceed from partial flexion in *A* to full extension in *C*. As the pattern is reversed from sitting to hyperflexion, the knees are again inclined to flex.

If total pattern were continued from *C,* the supine

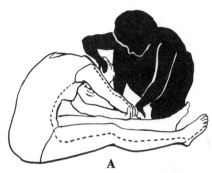

A

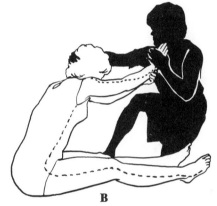

B

C

position would be assumed. To prevent further extension and to maintain sitting balance, head and neck and trunk flexor patterns must interact with extensor patterns.

MAT ACTIVITIES
Sitting

Fig. 66. Rising to sitting from supine.

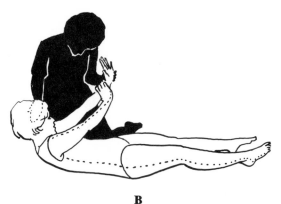

A

B

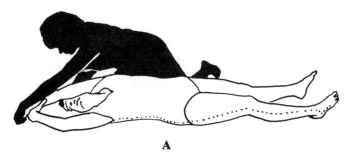

C

Component patterns

RESISTED

Head and neck, flexion with rotation to left
Upper trunk, flexion with rotation to left
Upper extremities, asymmetrical extension to left
(chopping)

FREE

Lower extremities adjust with bilateral flexion of hips

A. Lengthened range

COMMANDS

"Pull your arms down, lift your head toward me, and sit up!"

SUGGESTED TECHNIQUES

Traction to upper extremities, stretch and resistance

B. Approaching middle range

COMMANDS

"Push your arms down! Hold it!"

SUGGESTED TECHNIQUES

Repeated contractions, slow reversal

C. Shortened range

COMMANDS

"Hold it! Don't let me push you back!"

SUGGESTED TECHNIQUES

Approximation to upper extremities, rhythmic stabilization (see Balance—Sitting, Fig. 67)

Antagonistic pattern

Reversal to supine position

Note: Lower trunk and lower extremities will adjust with asymmetry toward the right in proportion to the extent of rotation of upper trunk to the left. The knees, extended in *A*, will incline toward flexion as shortened range of total pattern is approached, *C*.

If total pattern were continued, hyperflexion (Fig. 65, *A*) would be assumed. To achieve erect posture, and to maintain sitting balance, head and neck and upper trunk extensor patterns must interact with flexor patterns.

Configuration of head, neck, and trunk should be compared with configuration in Figure 65. Total movement from position *A*, Figure 65, to position *A*, this figure, and reversal to position *A*, Figure 65, are antagonistic movements of which erect sitting is approximately middle range.

Facilitation of total patterns

Fig. 67. Sitting balance.

Component patterns

A. RESISTED

Head and neck, flexion with rotation to left
Right upper extremity, extension–adduction–internal rotation, scapula depressed anteriorly

FREE

Left upper extremity, extension–abduction–internal rotation
Lower extremities adjust with asymmetrical flexion toward right

B. RESISTED

Upper trunk rotation toward right

FREE

Head and neck, mid-position with inclination toward rotation to left
Left upper extremity adjusts in extension with adduction, right in extension with abduction
Lower extremities adjust with asymmetrical flexion toward left

C. RESISTED

Upper extremities, reciprocal flexion–adduction–external rotation of left and extension–abduction–internal rotation of right

FREE

Head and neck, trunk and extremities adjust with symmetry

COMMANDS

A, "Hold! Don't let me push you back!"
B, "Hold! Don't let me turn you!"
C, "Hold, Don't let me move you!"

SUGGESTED TECHNIQUES

A, rhythmic stabilization of flexion, then extension components
B, rhythmic stabilization
C, rhythmic stabilization

A

B

C

Note: To promote sitting balance, antagonistic patterns are resisted alternately, with manual contacts shifted from anterior, *A,* to posterior; or at the same time, *B* and *C,* with anterior and posterior manual contacts. Whenever possible, approximation is applied at head, or shoulders, and is followed immediately by rhythmic stabilization. Equilibrium may be disturbed slowly without defeating ability to maintain position, or abruptly so that equilibrium must be recovered. Rocking movements against resistance should also be performed.

Other forms of sitting, side-sitting, long-sitting, sitting in a chair or on a table with feet free of support, and squat-sitting should be used. Resistance and rhythmic stabilization may be applied.

Fig. 68. Sitting—lower trunk rocking.

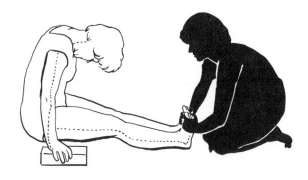

A

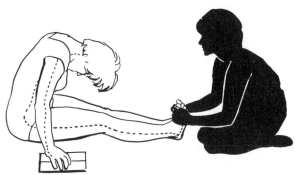

B

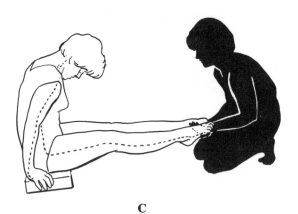

C

Component patterns

A. Approaching middle range—Rocking backward

RESISTED

Lower trunk, pulling backward

FREE

Upper extremities elevate and support trunk free of contact with surface
Head and neck, flexion

B. Shortened range—Rocking backward

RESISTED

Lower trunk, pulling backward

FREE

Upper extremities maintain trunk elevation free of contact with surface
Head and neck, flexion

C. Approaching middle range—Rocking forward

RESISTED

Lower trunk, thrusting forward

FREE

Upper extremities support trunk elevation and assist thrusting of lower trunk
Head and neck, flexion

Commands

"Push on your hands and lift yourself from the mat."
A. "Hold it!"
B. "Now, pull your hips back and away from me!"
C. "Push your hips toward me!"

Suggested techniques

Slow reversal, slow reversal—hold

Antagonistic pattern

Rocking forward is antagonistic to rocking backward, and the reverse

Note: Rocking movements promote balance in various postures. Here rocking of lower trunk backward and forward while body is free of surface contact promotes ability to perform transfer activities, as shown in Fig. 87, *B*. Lower extremities may be resisted alternately or reciprocally, if ability permits.

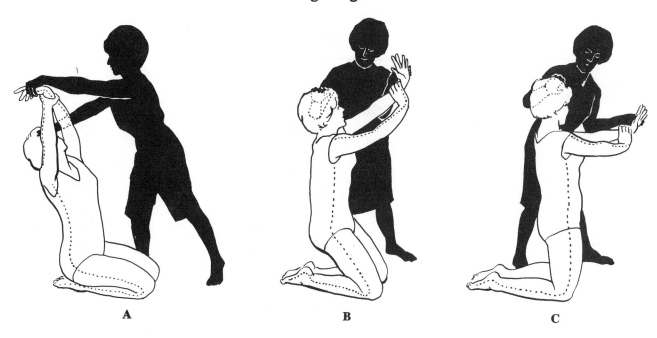

A   **B**   **C**

Fig. 69. Rising to kneeling toward left.

Component patterns

RESISTED

Head and neck, flexion with rotation to left
Upper extremities, bilateral asymmetrical extension to left (chopping)

FREE

Lower trunk adjusts in extension
Lower extremities adjust with hips extending, knees flexed

A. Initial position—
Heel-sitting, upper trunk extension to right

COMMANDS

"Pull your arms and your head down toward me, and get up on your knees!"

SUGGESTED TECHNIQUES

Stretch and resistance

B. Middle range

COMMANDS

"Up on your knees!" "Push with your head and arms!"

SUGGESTED TECHNIQUES

Maintain head and neck at middle range so that subject can push up onto knees

C. Shortened range

COMMANDS

"Hold it!" "Stay here!"

SUGGESTED TECHNIQUES

Rhythmic stabilization, slow reversal

Antagonistic pattern

Reversal to initial position with lifting

Note: This total pattern uses flexor patterns of upper trunk performed from lengthened range. In *B*, subject could flex elbows and pull to kneeling. For completely erect posture, upper trunk must shift toward extension.

Kneeling may be assumed from a position of total flexion, head on knees, with resistance applied at head, or at head and with upper extremities lifting; or, from heel sitting with trunk erect and resistance given at pelvis, or shoulders; or, subject may pull up on a stationary object (see Fig. 72). Rising with total extension is a more difficult and more advanced activity than is rising with flexion as shown in this figure.

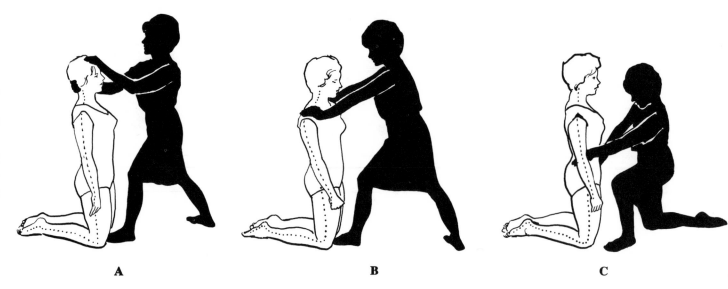

Fig. 70. Kneeling balance.

Component patterns

A. RESISTED

Head and neck, flexion and extension with rotation

FREE

Trunk and lower extremities adjust with flexion and extension patterns

Upper extremities adjust with compensatory movements if balance is disturbed

B. RESISTED

Upper trunk, rotation to left

FREE

Head and neck shift toward rotation to right

Upper trunk adjusts with flexion and extension toward left

Lower trunk adjusts in mid-posititon or toward right

Upper extremities adjust with compensatory movements, left in adduction, right in abduction

C. RESISTED

Lower trunk rotation to right

FREE

Head and neck and upper trunk shift toward rotation to left

Upper extremities adjust with compensatory movements as necessary

Lower extremities adjust toward right, extension and abduction of right, flexion and adduction of left

COMMANDS

A, "Hold everything! Don't let me move you!"
B, "Hold! Now, turn to the left, and hold!"
C, "Hold! Don't let me turn you!"

SUGGESTED TECHNIQUES

A, rhythmic stabilization
B, slow reversal—hold, rhythmic stabilization
C, approximation to right hip, rhythmic stabilization

Note: Other appropriate manual contacts include: head and shoulder, shoulder and pelvis on opposite sides. All manual contacts are combined for stability, one is anterior, the other posterior. Rocking movements against resistance should be used as well as abrupt but guarded disturbance of equilibrium.

Kneeling balance demands that knee flexors cooperate with hip extensors and that knee extensors interact with hip flexors. Thus, more advanced patterns of lower extremities are activated (see *Note*, Fig. 63).

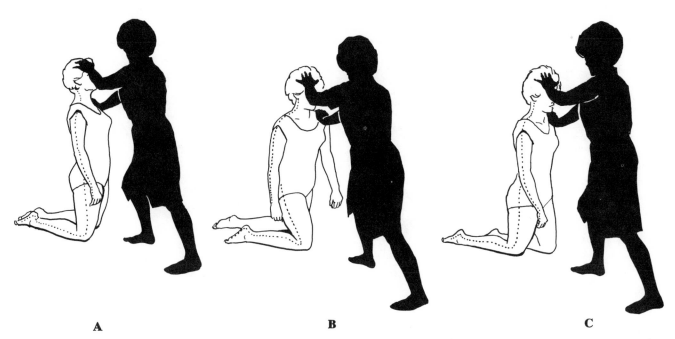

Fig. 71. Knee walking forward to right.

Component patterns

RESISTED

Head and neck, flexion with rotation to right
Upper trunk, flexion with rotation to right

FREE

Upper extremities adjust alternately between adduction and abduction
Right lower extremity, flexion–abduction–internal
rotation
Left lower extremity, flexion–adduction–external rotation

A. Initial position

COMMANDS

"Pull your head and left shoulder toward me, and
hold!"
"Now move your right knee forward toward me!"

SUGGESTED TECHNIQUES

Resistance

B. Shortened range—Right lower extremity

COMMANDS

"Put your weight on your right knee!"

SUGGESTED TECHNIQUES

Maintained resistance at head and shoulder

C. Approaching shortened range—
Left lower extremity

COMMANDS

"Pull your left knee toward the right and put your
weight on it."

SUGGESTED TECHNIQUES

Maintained resistance at head and shoulder

Antagonistic pattern

Knee walking backward to left

Note: Other appropriate manual contacts include:
head alone, shoulders, shoulder and pelvis on opposite
sides, and pelvis with therapist kneeling or half-kneeling (see Fig. 70).

Developmentally, the child does not walk on his
knees before performing various patterns of bipedal
walking. Nevertheless, normal children and adults use
knee walking when it is the most convenient mode of
progression. Knee walking may be used to promote
stability for bipedal walking.

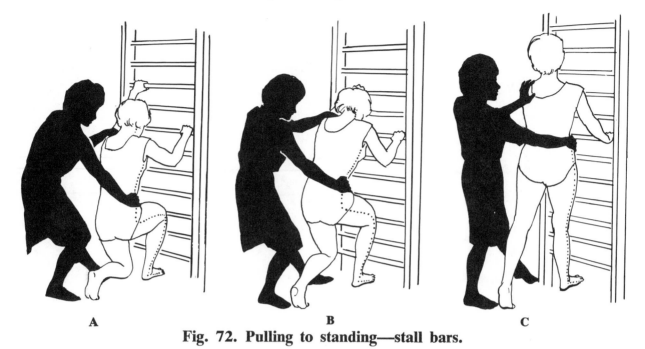

A   B   C

Fig. 72. Pulling to standing—stall bars.

Component patterns

RESISTED

Upper trunk, flexion with rotation to right, extension with rotation to left

Right lower extremity, extension–abduction–internal rotation

FREE

Head and neck adjust from flexion toward left to extension toward right

Left upper extremity extends toward adduction with elbow flexing, then extending

Right upper extremity extends toward abduction with elbow flexing, then extending

Left lower extremity, extension–adduction–external rotation

A. Initial position—Half-kneeling

COMMANDS

"Pull yourself forward, and push with your right foot!"

SUGGESTED TECHNIQUES

Stretch and resistance

B. Middle range

COMMANDS

"Now, push with both feet!" "Lift your head to the left!" "Stand up!"

SUGGESTED TECHNIQUES

Approximation at shoulder and hip, resistance

C. Shortened range

COMMANDS

"Hold!" "Now, pull your left foot forward and step on it."

SUGGESTED TECHNIQUES

Rhythmic stabilization, slow reversal

Antagonistic pattern

Reversal to half-kneeling

Note: Assumption of erect posture from half-kneeling, squat-sitting, or sitting in a chair requires that the total pattern be initiated with flexion followed by extension to upright. Head and neck lead the movements. Pulling to standing from sitting is shown in Figure 84.

Other activities to be performed at stall bars include: rising to kneeling (see Fig. 69), and climbing activities. Climbing may be considered upright quadrupedal locomotion. Lower extremity patterns may be emphasized while subject maintains position with upper and the opposite lower extremities (see Fig. 60).

MAT ACTIVITIES
Bipedal Progression

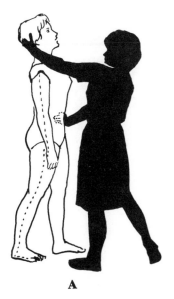
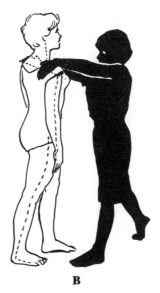
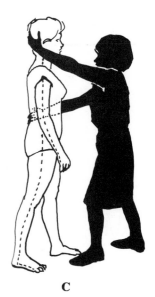

A   B   C

Fig. 73. Standing balance, stability.

Component patterns

A. RESISTED

Head and neck, extension with rotation to right
Lower trunk, rotation to right

FREE

Right lower extremity adjusts with weight-bearing
Left lower extremity adjusts toward flexion and swing phase
Upper extremities adjust with compensatory movements

B. RESISTED

Upper trunk, rotation to right

FREE

Head and neck shift toward rotation to left
Upper extremities adjust with compensatory movements
Right lower extremity adjusts in extension–abduction–internal rotation
Left lower extremity adjusts in flexion–adduction–external rotation

C. RESISTED

Head and neck, extension with rotation to right
Lower trunk, rotation toward left

FREE

Upper extremities adjust with compensatory movements

Right lower extremity, extension–abduction–internal rotation
Left lower extremity, extension–adduction–external rotation

COMMANDS

A, "Hold! Don't let me pull your head forward! Don't let me push your hip back!"
B, "Hold! Don't let me turn you to the left!"
C, "Hold! Don't let me pull you forward!"

SUGGESTED TECHNIQUES

A, maintain resistance at head, increase pressure at hip
B, rhythmic stabilization, slow reversal—hold
C, rhythmic stabilization, approximation on right

Note: In *A* and *B*, balance is stable because one manual contact is anterior, the other is posterior. In *C*, both contacts are posterior. As subject is pulled forward to disturb balance, response of extension patterns of lower extremities is demanded.

Other appropriate manual contacts include: head alone, shoulder and pelvis at opposite sides, and pelvis on both sides. See also, Figure 74.

Standing may be assumed from half-kneeling, squat-sitting, sitting in a chair, and from plantigrade. Except for the plantigrade position where trunk is flexed, all assumptions of standing have an initial phase of flexion of head and neck, and of upper trunk, followed by the extension phase. See Figure 84 for rising to standing from sitting in wheel chair.

Use of the developmental sequence   **159**

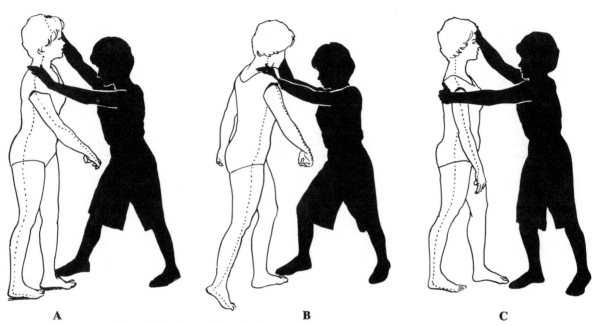

Fig. 74. Standing balance, compensatory movements.

Component patterns

A. RESISTED

Head and neck, flexion with rotation to left
Upper trunk, flexion with rotation to left

FREE

Upper extremities adjust with compensatory movements; right, extension–adduction–internal rotation; left, extension–abduction–internal rotation

Left lower extremity, flexion–abduction–internal rotation

Right lower extremity, flexion–adduction–external rotation

B. RESISTED

As in *A*

FREE

Upper extremities, as in *A*

Left lower extremity, extension–adduction–external rotation

Right lower extremity, extension–abduction–internal rotation

C. RESISTED

Head and neck, flexion with rotation to left
Upper trunk, rotation to right

FREE

Upper extremities adjust with compensatory movements

Lower extremities adjust alternately between flexion patterns, *A*, and extension patterns, *B*

COMMANDS

A, "Hold, don't let me push you back!"
B, "Come back to me!"
C, "Now, stand tall, and hold!"

SUGGESTED TECHNIQUES

A, pressure and stretch
B, resistance
C, rhythmic stabilization, alternate increase of pressure at head, then at shoulder

Note: Whereas in Figure 73, subject was encouraged to remain stable, in this instance balance has been disturbed to elicit compensatory movements through range, *A* and *B*. In *C*, subject is again stabilized, one manual contact is anterior, the other, posterior.

Rocking movements should be performed against resistance. Abrupt disturbance of equilibrium may be done using a variety of manual contacts. For completeness, positions of feet should be reversed; symmetrical positions should be used as well.

MAT ACTIVITIES
Bipedal Progression

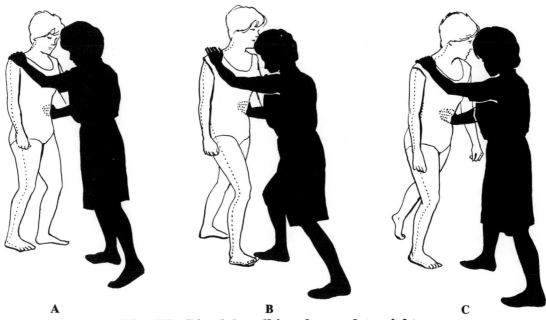

A   B   C

Fig. 75. Bipedal walking forward to right.

Component patterns (see Note).

RESISTED

Upper trunk, rotation to left
Lower trunk, rotation to right

FREE

Upper extremities adjust with compensatory movements

Right lower extremity alternates between flexion–abduction–internal rotation and extension–adduction–external rotation

Left lower extremity alternates between extension–abduction–internal rotation and flexion–adduction–external rotation

A. Initial position

COMMANDS

"Lift your right foot forward to the right."

SUGGESTED TECHNIQUES

Approximation on left, resistance on left

B. Heel strike, Right lower extremity; Stance phase on left

COMMANDS

"Step on your right foot, and push with your left."

SUGGESTED TECHNIQUES

Resistance

C. Stance phase, Right lower extremity; Preparatory for swing phase of left

COMMANDS

"Pull your left foot forward!"

SUGGESTED TECHNIQUES

Approximation to right, resistance on left

Antagonistic pattern

Bipedal walking backward to left

Note: Therapist and subject have adjusted to needs of photography. Therapist should be positioned toward right. Because therapist is not in the diagonal position, subject has not moved in a diagonal direction. Although instructed to walk toward the right, subject proceeds to move toward therapist. This figure illustrates the importance of the therapist's assuming a position on the diagonal if a diagonal direction or movement is expected. Other distortions include: *B,* failure of subject's right upper extremity to extend toward abduction although left upper extremity has flexed toward adduction, incomplete eversion of left foot; *C,* excessive leaning of subject toward therapist with failure of head to shift toward right lower extremity in approaching stance phase.

Manual contacts for alternating resistance to swing phase with approximation on stance phase should be shifted to level of pelvis bilaterally, as for therapist's right hand.

GAIT ACTIVITIES

The ability to walk erect materializes from the performance of less advanced activities. When methods of facilitation are used to hasten motor learning, the ability to walk may be enhanced by intensive performance of less advanced preparatory activities. Thus, the training of gait patterns begins with rolling and proceeds with prone locomotion on a flat surface provided by a gymnasium mat. Assuming sitting, quadrupedal, kneeling, and standing postures, too, are preliminaries for walking in the upright position. The ability to roll, creep, and stand does not insure the ability to walk; but, the quality of a gait pattern may be improved and the patient's potentials may be more fully developed by intensive performance of less advanced activities.

Assuming an erect posture and walking are goals for the majority of patients. Walking may be an unattainable goal for severely disabled patients. Some patients may achieve quadrupedal upright walking (two crutches or two canes) or tripedal walking (one crutch or one cane). Others may only acquire the ability to cruise in a walker. Those with gross deficiencies may be limited to manual locomotion in a wheel chair. Nevertheless, the urge to walk is basic to the human species. Maximal effort should be made to achieve the highest level of ability that has functional significance for the patient.

The developmental activities related to locomotion in the upright position are shown in Table 2 (see pp. 120–121). The overlapping between mat activities and gait activities is evident. The least advanced form of assuming erect posture is that of "Pulling to standing from sitting." Using parallel bars for pulling to standing is portrayed as a wheel chair activity. Patterns and techniques of proprioceptive neuromuscular facilitation are used in gait activities with all type of patients and with various kinds of support, such as parallel bars, braces, crutches, and canes.

Standing Balance

Having assumed an erect position, standing balance is necessary to maintenance of posture. Maintenance of erect posture and postural adjustments during bipedal activity are dependent upon postural and righting reflexes and a neat interaction between antagonistic pairs of muscle groups.

When the normal subject attempts to maintain himself in an erect position with feet firmly planted, and when, for example, his balance is disturbed by pushing against his forehead, the dorsiflexors contract in an effort to maintain the position of the feet and ankles. As resistance is increased, the antagonistic muscles, the plantar flexors, contract in cooperation with the dorisflexors. The effort to maintain position is expressed in isometric contraction of the responsible muscle groups and a co-contraction of antagonists may result. If balance is disturbed sufficiently, the plantar flexors, by isotonic contraction, contribute to recovery of the position through movement. Furthermore, a whole complex of muscle groups reinforces the effort. Related muscles of the neck, trunk, and the extremities respond. Those responses which entail dorsiflexion of the foot and ankle prepare for the swing phase of walking; responses of plantar flexion prepare for the stance phase and propulsion.

To stimulate postural responses and to facilitate response of specific groups of muscles, patterns and techniques of proprioceptive neuromuscular facilitation are used in the training of standing balance, as shown in mat activities. Pressure and resistance, maximal for the occasion, are applied through specific manual contacts. Other techniques, including rhythmic stabilization, repeated contractions, reversal of antagonists, and approximation may be superimposed. Selectivity of response is achieved by disturbing balance in a diagonal direction.

Just as the specific patterns of facilitation are composed of two diagonals of movement which, in turn, consist of two pairs of antagonistic patterns, so are there two diagonals and two pairs of antagonistic patterns within the total pattern of erect posture. Balance may be disturbed by applying pressure at the head, shoulder girdle, or pelvis in a direction which is from left anterior to right posterior, or in the reverse direction, from right posterior to left anterior. In this way, one pair of antagonistic patterns or the patterns of one diagonal may be stimulated. If pressure is applied in a direction which extends from right anterior to left posterior, or in the reverse direction from left posterior to right anterior, the second pair of antagonistic patterns of the second diagonal responds.

Rotation components within the neck and trunk and the extremities may be used to promote security and balance in the erect posture. Various combinations of extremity positions may be used as in a symmetric posture or reciprocation. To facilitate the response of rotation components, pressure is applied in an anterior-posterior direction on one side of the body and, at the same time, in a posterior-anterior direction on the opposite side of the body.

The significant aspects identified for the illustrations of mat activities apply to illustrations of gait activities.

The training of standing balance without use of supportive devices has been portrayed as a mat activity in Figures 73 and 74. Training of balance in the upright posture may be carried out while using support as shown for parallel bars, Figures 76 and 77; for crutches, Figures 78 and 79.

In the developmental sequence there is an interweaving of movement and balanced posture. For development of gait or walking patterns, balancing activities are interspersed with total movement patterns related to gait. In this way, the patient is helped to walk or to ambulate as independently as possible.

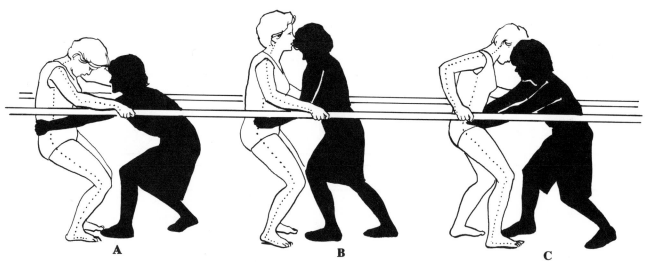

Fig. 76. Sitting to standing balance.

Component patterns

A. RESISTED

Approaching symmetrical total flexion as in lowering from standing to sitting in a chair

FREE

Head and neck, trunk, and lower extremities adjust in flexion

Upper extremities prevent lowering to squat-sitting

B. RESISTED

Approaching symmetrical total extension as in rising from sitting to standing

FREE

Head and neck, trunk, and lower extremities adjust toward extension

Upper extremities prevent lowering toward sitting

C. RESISTED

Approaching total extension with extremities in diagonal reciprocation

FREE

Head and neck, and trunk adjust in flexion in effort to align superior region with support on feet

Left upper extremity pulls as right pushes to assist alignment of superior region

COMMANDS

A, "Hold! Don't let me pull you forward!"

B, "Hold! Don't let me push you back! Now, pull with your arms, and come on up toward me!"

C, "Hold!" "Now, push up to me, and lift your head up!"

SUGGESTED TECHNIQUES

A and *B*, slow reversal—hold with rocking backward and forward, rhythmic stabilization *C*, rhythmic stabilization

Note: Subject has not approached the shortened range of extension of lower extremities. Therefore, approximation at hips cannot be used to stimulate extensor reflexes. In *C*, manual contact for head and neck extension would promote total extension.

The ability to maintain a semi-flexed posture is useful in the initial phase of rising to standing from sitting, and in lowering to sitting in a chair. Resistance applied at various ranges of a total pattern hastens the ability to move through increased range. If assistance to upright is given, rocking movements should be used to promote control and the ability for rising independently. Pulling to standing in parallel bars is shown in Figure 84.

Balancing activities in an erect posture are shown in Figure 77. Figures 73 and 74 portray training of standing balance in open area.

GAIT ACTIVITIES
Parallel Bars

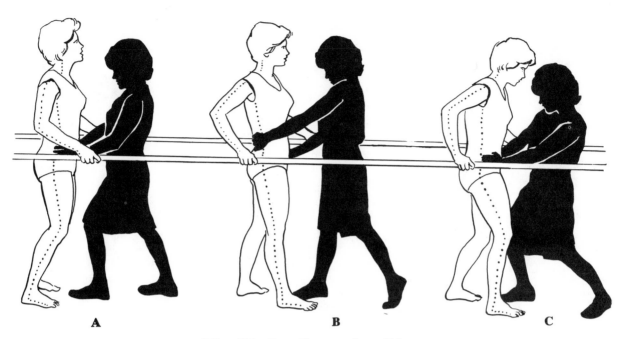

Fig. 77. Standing and walking.

Component patterns

A. RESISTED

Lower trunk, rotation to left

FREE

Head and neck shift toward left as extension of left lower extremity is activated

Right lower extremity has flexed prior to extending

Upper extremities adjust with stabilization

B. RESISTED

Lower trunk, rotation to right

FREE

Head and neck adjust toward left, away from advanced right lower extremity

Extremities, positioned in diagonal reciprocation, adjust with stabilization

C. RESISTED

Right lower extremity, approaching stance phase

Left lower extremity, approaching swing phase

FREE

Head and neck adjust toward left

Upper trunk adjusts toward right

Lower trunk adjusts toward left

Left upper extremity pulls to assist swing of right lower extremity; right upper pushes to assist propulsion of left lower

COMMANDS

A and *B*, "Hold! Don't let me turn you!"

C, "Step on your right foot, and move your right hand forward. Now, pull your left foot forward."

SUGGESTED TECHNIQUES

A and *B*, rhythmic stabilization, approximation

C, approximation on right, resistance on left

Note: In *A* and *C*, approximation is needed to promote extension of weight-bearing lower extremities.

For other appropriate manual contacts, see Figures 73–75.

Use of the developmental sequence   165

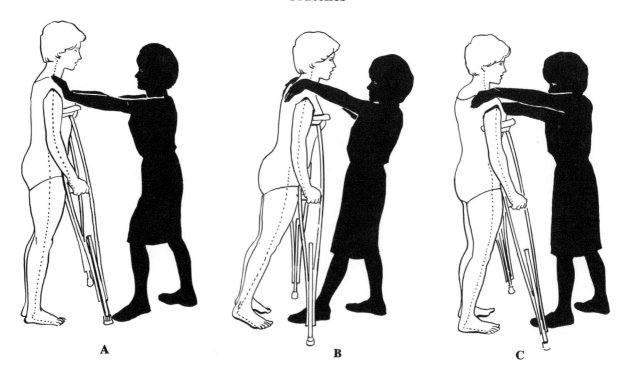

Fig. 78. Superior region balance.

Component patterns

A. RESISTED

Upper trunk, flexion

FREE

Head and neck adjust with flexion
Upper extremities thrust downward and backward
Lower trunk adjusts with flexion
Lower extremities adjust with activation of dorsi-flexors

B. RESISTED

Upper trunk, extension

FREE

Head and neck adjust with extension
Upper extremities thrust downward and forward
Lower trunk adjusts with extension
Lower extremities adjust with activation of plantar flexors

C. RESISTED

Upper trunk, rotation toward right

FREE

Head and neck adjust toward right

Lower trunk adjusts with rotation toward left
Upper and lower extremities adjust with stabilization

COMMANDS

A, "Hold! Don't let me push you back!"
B, "Hold! Don't let me pull you forward!"
C, "Hold! Don't let me turn you to the left!"

SUGGESTED TECHNIQUES

A and *B*, pressure and resistance
C, approximation to right shoulder, rhythmic stabilization

Note: In *A* and *B*, stability is threatened so that compensatory movements may be necessary to recover equilibrium. In *C*, stability is encouraged through resistance applied simultaneously to flexion and extension of trunk. Other appropriate manual contacts include head, and head and shoulder on opposite sides.

Resistance may be given at subject's wrist as a crutch is shifted in position, and as the crutch is maintained in position. Subject may be instructed to hold one crutch free of contact with surface as resistance is applied in various combinations. Various combinations of foot and crutch positions should be used. Rocking movements may be done with crutches lifted as subject rocks backward and replaced in contact as subject rocks forward.

GAIT ACTIVITIES
Crutches

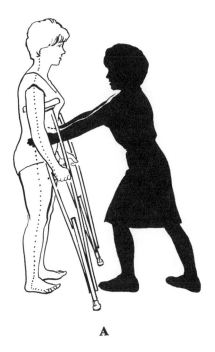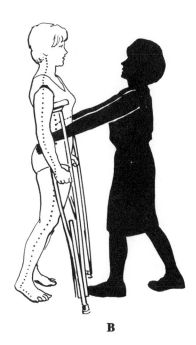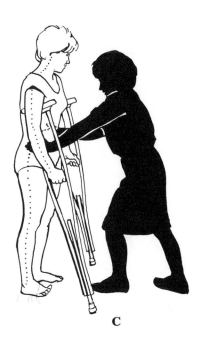

A   B   C

Fig. 79. Inferior region balance.

Component patterns

A. RESISTED

Lower trunk, flexion

FREE

Head and neck, and upper trunk adjust with flexion
Upper extremities thrust downward and backward
Lower extremities adjust with activation of dorsiflexors

B. RESISTED

Lower trunk, extension

FREE

Head and neck, and upper trunk adjust with extension
Upper extremities thrust downward and forward
Lower extremities adjust with activation of plantar flexors

C. RESISTED

Lower trunk, rotation toward left

FREE

Head and neck adjust toward left
Upper trunk adjusts with rotation toward right
Upper and lower extremities adjust with stabilization

COMMANDS

A, "Hold! Don't let me push you back!"
B, "Hold! Don't let me pull you forward!"
C, "Hold! Don't let me turn you to the right!"

SUGGESTED TECHNIQUES

A and B, pressure and resistance
C, approximation at left hip, rhythmic stabilization

Note: In A and B, as in Figure 78, stability is threatened. In C, stability is encouraged. Various combinations of foot and crutch positions, one crutch free of contact, and one foot free of contact with surface may be used.

During balancing activities in the erect posture, whenever extension of hips and knees is incomplete at the appropriate time, approximation should be used to promote response of postural reflexes, except where skeletal structures are not intact.

Balance in any position may be disturbed abruptly with appropriate guarding for safety. Balancing in side-lying, on elbows and knees, on hands and knees, on hands and feet, and in various forms of sitting contribute to the development of standing balance. Assumption of various postures also contribute to assumption of standing and to maintenance of upright posture.

Use of the developmental sequence   **167**

Walking

Normal gait involves smooth, rhythmical, and continuous transition among component patterns of the total movement. Even though walking proceeds from a swing phase (flexion) through a stance phase (extension), and the movement of one lower extremity is timed with alternating phases of the other lower extremity, all components of motion within the neck and trunk and extremities are used as necessary.

Walking against resistance in a diagonal direction may be regarded in much the same way as is standing balance when disturbed and recovered in a diagonal direction. In walking there is a transition from balance or sustained posture to movement. Whereas in balancing activities effort and movement are directed centrally toward a point of balance, bipedal progression requires continuous effort in the direction of the total movement. The first phase of response that occurs when balance is disturbed in an anterior-posterior direction is related to the swing phase of walking forward; the phase of recovery of balance is related to the stance and propulsive phase. Thus, patterns of movement necessary to walking are developed during standing balance activities.

During standing balance activities the emphasis is on stability of segments promoted by isometric contractions and co-contraction of antagonistic muscle groups, but movement with isotonic contraction of muscle group supports the effort to recover balance. In walking against resistance, the emphasis is on movement of the segments promoted by isotonic contractions of muscle groups, but balance and posture with isometric contractions of related muscle groups support the effort to move. When in standing balance a co-contraction of antagonists is promoted by maximal resistance there is no movement; the segments of the body unite to form a stable pillar. When the effort to walk is resisted, the segments of the body interact in unison and depart toward a goal.

A normal person walking into a gale of fifty miles per hour leans forward as he walks. The flexion component dominates the movement. If he turns and walks backward into the wind, he extends his neck and trunk so that the extension component dominates the movement. When manual resistance is used the result is much the same. In walking forward, the tendency is to lean into the resisting force. When walking backward, the tendency is the same—to lean into the resisting force. By so doing, the propulsive effort becomes more effective. To encourage the maintenance of an upright posture and to discourage the patient's dependence on the therapist for support, resistance must be adjusted for control of the head and neck and trunk. If, for example, related component patterns of the head and neck and shoulder are resisted, the patient is resisted strongly and then is required to "hold" or maintain the position of his head and shoulder as he moves his extremities. In this way he learns to control his entire body as he moves.

The direction of the diagonals, the pairs of antagonistic patterns, the direction of pressure and resistance, and the manual contacts used are the same for resisted walking as for standing balance. The therapist's approach is the same; the therapist must accommodate to the patient's effort and movement.

ILLUSTRATIONS

Specific comments on illustrations and legends which follow the discussion on mat activities apply to those which portray walking and the use of stairs.

Bipedal walking without support has been portrayed as a mat activity, Figure 75. Walking while using support of parallel bars has been shown in Figure 77C. Use of crutches is portrayed in Figure 80; and ascent and descent of stairs are shown in Figures 81 and 82.

As a patient walks, his gait pattern may deteriorate because of fatigue, imbalances between antagonistic patterns or muscle groups, or pain. The intermittent use of balancing activities with rhythmic stabilization applied to the segment of concern may help to restore proper interaction of segments and antagonistic muscle groups. A change in direction may alleviate fatigue.

To firmly establish a proper gait pattern, use of the pattern is necessary. If additional support in the form of braces or crutches permits a patient to use a desirable pattern for longer periods of time, support should be used. Gripping parallel bars or crutches during resisted balance and walking provides additional reinforcement. Supportive devices should be viewed as tools which permit the patient to put forth greater effort for longer periods in good form. Although some patients may need supportive devices permanently, other patients may abandon support sooner by intensive use against resistance. While braces limit movement and while they provide security, the proximal to distal effect of resisted balancing and walking activities permits response of related patterns and muscle groups insofar as the potential for response is present. The influence of postural and righting responses dominates.

Resistance should be applied so as to give security or challenge the patient according to his needs. The task of the physical therapist is to provide the patient with an opportunity to improve and to progress. Wisely, challenge should be interspersed with security to help the patient toward independence.

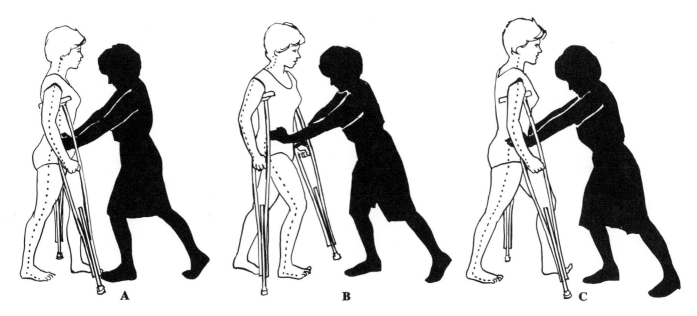

Fig. 80. Walking forward.

Component patterns

RESISTED

Lower extremities, alternating swing and stance phases

FREE

Head and neck adjust toward extremity in stance phase
Upper trunk adjusts toward extremity in swing phase
Lower trunk adjusts toward extremity in stance phase
Upper extremities reciprocate with lower extremities

A. Initial position

COMMANDS

"Move your left crutch, then your right foot forward."

SUGGESTED TECHNIQUES

Approximation on left, resistance on right

B. Approaching stance phase on right, Propulsion and swing on left

COMMANDS

"Step on your right foot, move your right crutch and your left foot forward."

SUGGESTED TECHNIQUES

Resistance on left, approximation on right

C. Approaching heel strike on left, Propulsion on right

COMMANDS

"Step on your left foot, and move your left crutch forward."

SUGGESTED TECHNIQUES

Approximation on right, resistance on left

Antagonistic Pattern

Walking backward

Note: Subject is walking with diagonal reciprocation using a four-point gait. Other crutch-gait patterns may be trained with appropriate adjustments of the therapist's position, manual contacts, and commands. Walking forward and backward emphasize flexion and extension components. Side-stepping, turning or pivoting, walking diagonally forward and backward toward left and right should be used.

The use of crutches while ascending and descending ramps and stairs, rising to standing and lowering to sitting in a chair may be trained in a similar fashion. Safety is always of primary concern. Training by use of preparatory activities of the developmental sequence lessens the hazards of locomotion in the upright posture.

The use of other types of crutches, of a cane, or of canes, may be taught with proper adaptations according to the type of support.

GAIT ACTIVITIES
Stairs

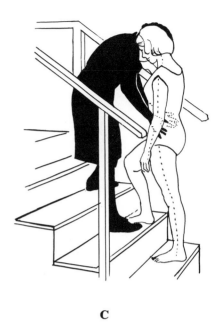

A   B   C

Fig. 81. Ascending forward.

Component patterns

RESISTED

Lower extremities, alternating flexion and extension

FREE

Head and neck adjust toward extending lower extremity

Right upper extremity pulls on hand rail to assist left lower extremity

Left upper extremity reciprocates with right lower extremity

Upper trunk adjusts toward flexing lower extremity

Lower trunk adjusts toward extending lower extremity

A. Initial position

COMMANDS

"Hold! Now, step up with your left foot."

SUGGESTED TECHNIQUES

Approximation at pelvis, resistance on left

B. Approaching extension on left, Flexion on right

COMMANDS

"Push with your left foot, and step up with your right."

SUGGESTED TECHNIQUES

Approximation on left, resistance on right

C. Approaching flexion on left, Extension on right

COMMANDS

"Push with your right foot, and step up with your left."

SUGGESTED TECHNIQUES

Approximation on right, resistance on left

Antagonistic Pattern

Descending backward

Note: Balancing activities with rhythmic stabilization may be used. Manual contacts for head and pelvis, shoulder and pelvis may be used. By taking a position behind subject, therapist may resist flexion patterns of lower extremity as shown for creeping forward, Figure 61; if possible, both hand rails may be grasped, or both hands may grasp one hand rail. When necessary, one lower extremity may lead the total pattern repeatedly while the other is brought to the level of the first rather than advancing to the next step.

Ascending forward and descending backward may be done as a plantigrade activity. Ascending backward and descending forward may be done while sitting. Other activities include balancing while using crutches or canes, and ascending and descending while using supportive devices. Ascending and descending, a ramp may be considered as preparatory to ascending and descending stairs.

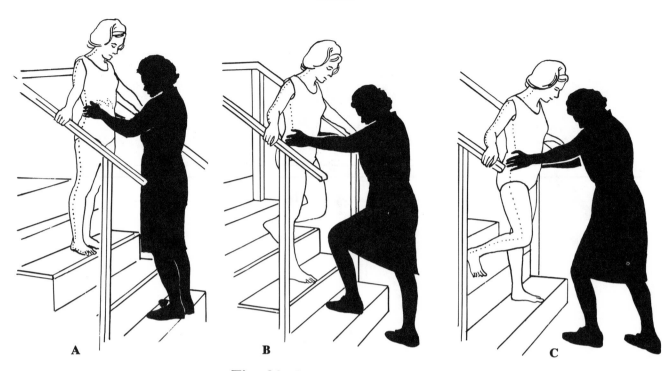

Fig. 82. Descending forward.

Component patterns

RESISTED

Lower extremities, alternating flexion and release of extension

FREE

Head and neck adjust toward extended lower extremity

Upper extremities support releasing lower extremity

Upper trunk adjusts toward extension and releasing lower extremity

Lower trunk adjusts toward extension and extended lower extremity

A. Initial position

COMMANDS

"Hold! Now reach down with your right foot."

SUGGESTED TECHNIQUES

Approximation, left and right, followed by resistance on right

B. Approaching flexion on left, Release of extension on right

COMMANDS

"Reach with your left foot, and bend your right knee slowly!"

SUGGESTED TECHNIQUES

Approximation on right, resistance on left

C. Approaching flexion on right, Release of extension on left

COMMANDS

"Now hold back with your left knee, and reach down with your right foot."

SUGGESTED TECHNIQUES

Approximation on left, resistance on right

Antagonistic Pattern

Ascending backward

Note: In *B*, left lower extremity is about to advance forward and downward with flexion of hip and extension of knee as right lower extremity enters phase of releasing extension of hip and knee, as shown in *C*.

Balancing activities should include advancing lower extremity free of contact as well as rhythmic stabilization done during various phases of the total pattern.

WHEEL CHAIR AND TRANSFER ACTIVITIES

As the normal child grows, he is often provided with toys which promote, to a degree, his motor development. The child who moves a toy with foot pedals or by cooperative effort of upper and lower extremities is developing movement patterns necessary to balanced posture and walking. Unless the child is handicapped by physical disability, he does not use a wheel chair. Yet, a wheel chair may be used by a child or an adult as a means of locomotion. The proper use of a wheel chair may further the recovery of the patient.

In the adapted developmental sequence, certain activities are closely related to wheel chair activities. Assuming and maintaining a sitting position, rising to standing from sitting, lowering to a squat position and sitting, and foot lifting and stamping prepare the patient to use a wheel chair. Bilateral movements of the upper extremities usually provide the force necessary to wheel a chair. However, ipsilateral use of one upper and one lower extremity may be the best means of propulsion for some patients. Again, the performance of related mat activities prepares for more complex functional activities.

Wheeling a chair demands coordination of body segments in balance and in movement. The ability to sustain and to recover a sitting posture is necessary to the patient's safety. The ability to move extremities while sitting permits the patient to propel the chair and to use it efficiently. Manipulation of brakes, foot plates, and wheels are necessary to operation. Entering, sitting, and arising from the chair are goals for the majority of patients. Learning to use a wheel chair skillfully may be hastened by application of proprioceptive neuromuscular facilitation.

As with other activities, using a wheel chair may be regarded as a total pattern of movement made up of component patterns. Resistance may be applied to the total pattern by restricting movement of the chair itself. In this way, some patients may improve their strength and increase their speed of performance.

In promoting the ability to transfer from chair to bed or table, or to and from an automobile, the individual patient's potentials and needs determine the method to be used and the selection of component patterns to be trained. Techniques of facilitation, including maximal resistance, approximation, reversal of antagonists, repeated contractions, rhythmic stabilization, and relaxation, may be used in training. Techniques are superimposed upon patterns of facilitation insofar as specific patterns contribute to training and upon functional movements, such as reaching for and lifting foot plates, placement of a transfer board, or removal of an arm of a chair.

ILLUSTRATIONS

An example of training of a specific component pattern necessary to the management of a wheel chair is shown in Figure 83. Pulling to standing at parallel bars, and transfer to bed and automobile are shown in Figures 84–87, while transfer to toilet is shown as a self-care activity, Figure 88.

The guidelines which accompany the illustrations and legends for mat activities apply to all subsequent illustrations and legends.

Training in the use of a wheel chair is viewed as a phase of the patient's total treatment program. Just as related mat activities prepare the patient to use a wheel chair, so do properly performed wheel chair activities prepare the patient for more advanced activities, such as household tasks. There is an overlapping between various phases of treatment with emphasis on those activities which will hasten the achievement of goals.

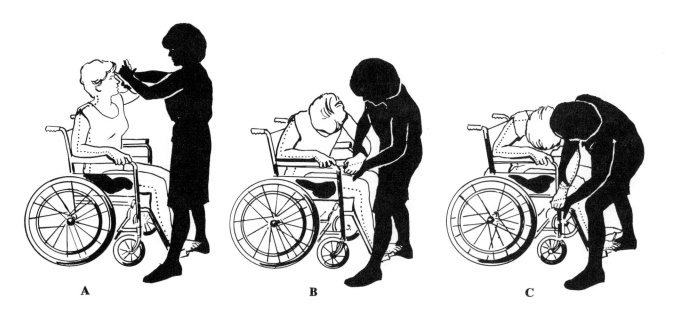

Fig. 83. Use of hand brake.

Component patterns

RESISTED

Left upper extremity, extension–adduction–internal rotation

FREE

Head and neck, and upper trunk, flexion with rotation to right

Right upper extremity grasps arm of chair for security and so as to reinforce left upper extremity

Lower trunk and lower extremities adjust for stability

A. Lengthened range

COMMANDS

"Squeeze my hand, and reach down and across to the brake."

SUGGESTED TECHNIQUES

Stretch and resistance

B. Middle range

COMMANDS

"Hold it! Now, reach again!"

SUGGESTED TECHNIQUES

Repeated contractions, slow reversal

C. Shortened range

COMMANDS

"Now, open your hand and grasp the brake." "Don't let me pull you away from it."

SUGGESTED TECHNIQUES

Resistance, repeated contractions

Antagonistic Pattern

Rising to sitting toward left

Note: Radial extensor thrust may be used with hand opening rather than closing. As shown, the goal is to train the ability to flex the upper trunk with rotation. If thrusting is used, inclination to extend trunk may occur.

In *C*, opening and closing of the hand may be performed against resistance. The effort to unlock and lock the brake may be resisted. The entire pattern of movement from *C* to *A* may be resisted with reversal of antagonist techniques. Other appropriate manual contacts include head and left upper extremity at wrist.

The same pattern with reversal may be used in training the ability to reach for the right knee, and if necessary, to assist the lifting of the right lower extremity away from and placement on foot plate. The manipulation of foot plates may be trained in a similar way.

WHEELCHAIR AND TRANSFER ACTIVITIES

Fig. 84. Pulling to standing.

Component patterns

RESISTED

Lower trunk and lower extremities, extension

FREE

Head and neck, and upper trunk adjust in flexion, then extension

Upper extremities grasp and pull on bars with elbows flexing, then extending

A. Initial position

COMMANDS

"Pull on the bars, pull toward me!"

SUGGESTED TECHNIQUES

Approximation, resistance

B. Middle range

COMMANDS

"Lift your head, look up here, and push!"

SUGGESTED TECHNIQUES

Resistance, repeated contractions, slow reversal

C. Approaching shortened range

COMMANDS

"Hold it! Now, push up all the way!"

SUGGESTED TECHNIQUES

Approximation, rhythmic stabilization

Antagonistic Pattern

Reversal to sitting

Note: Pulling to standing is the least advanced form of rising to standing. Therapist places right knee against subject's right knee (as for patient's less involved side) to promote stability.

Pushing to standing with subject's hands on arms of chair may be done by having subject extend head and neck, and upper trunk from a hyper-flexed position with head near left or right knee. Extension then proceeds toward the opposite side, right or left. Manual contacts with head and pelvis on opposite sides may be used.

Subject may also use one hand on bar, the other on arm of chair. Manual contacts may be made at shoulder on side of hand on chair arm, and at pelvis on side of hand on bar.

Component patterns which may be resisted include: placement of feet with knees flexed, shifting forward in seat of chair by rocking from side to side, and reaching for parallel bars.

Use of the developmental sequence

Fig. 85. From chair to bed.

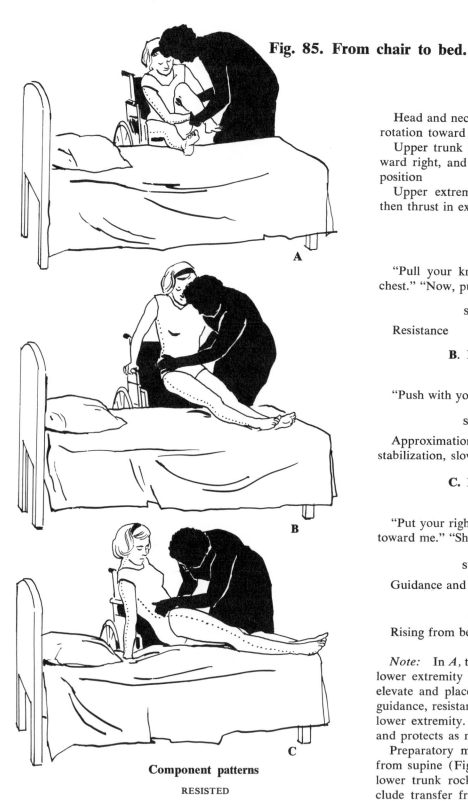

Component patterns

RESISTED

Upper extremities, placement of lower extremities on bed

Lower trunk, elevation and rotation toward left

FREE

Head and neck adjust with flexion, then flexion with rotation toward right

Upper trunk adjusts with flexion, then rotation toward right, and releases toward extension and supine position

Upper extremities flex to assist lower extremities, then thrust in extension for elevation of lower trunk

A. Initial position

COMMANDS

"Pull your knee away from me and toward your chest." "Now, push it forward to the bed."

SUGGESTED TECHNIQUES

Resistance

B. Elevation of lower trunk

COMMANDS

"Push with your hands and lift up!"

SUGGESTED TECHNIQUES

Approximation for initiation, resistance, rhythmic stabilization, slow reversal

C. Rotation of lower trunk

COMMANDS

"Put your right hand on the bed, and turn your hips toward me." "Shift your hand to the right, again!"

SUGGESTED TECHNIQUES

Guidance and resistance to lower trunk rotation

Antagonistic Pattern

Rising from bed to chair

Note: In *A,* therapist grasps subject's hands and left lower extremity and thereby resists subject's effort to elevate and place lower extremity. At the same time, guidance, resistance, or protection may be given to the lower extremity. In *B* and *C,* therapist guides, resists, and protects as necessary.

Preparatory mat activities include rising to sitting from supine (Fig. 66), sitting balance (Fig. 67), and lower trunk rocking (Fig. 68). Related activities include transfer from chair to bed, sideward approach with arm of chair removed; transfer from chair to mat platform; and transfer from chair to table using side approach with chair arms in place.

WHEELCHAIR AND TRANSFER ACTIVITIES

Fig. 86. From chair to standing to bed.

Component patterns

RESISTED

Lower trunk and lower extremities, extension with rotation toward left, then release to flexion toward left

FREE

Head and neck, and upper trunk adjust toward flexion with rotation toward right

Right upper extremity thrusts in extension

Left upper extremity does not contribute to total pattern

A. Initial position

COMMANDS

"Lean toward me, push with your right arm, and stand up!" "Straighten your knees!"

SUGGESTED TECHNIQUES

Approximation and resistance

B. Approaching shortened range, Standing

COMMANDS

"Hold it!" "Now, reach your right hand to the bed!"

SUGGESTED TECHNIQUES

Approximation at pelvis

C. Approaching shortened range, Sitting

COMMANDS

"Now, sit down slowly!"

SUGGESTED TECHNIQUES

Guidance and resistance to lower trunk

Antagonistic pattern

Rising from sitting to standing to chair (wheel chair placed at angle at foot of bed)

Note: Subject is using extremities of right side to achieve transfer. Therapist has placed right knee against subject's right knee to insure one pillar of support. Approximation and rotation of lower trunk toward left promotes stability of left lower extremity, *B*. Also in *B*, subject has achieved the maximum of extension which is limited by right upper extremity's contact with arm of chair. With approximation, freeing of right hand, and extension of head and neck, subject may achieve complete extension before turning and reaching for bed. Note that in transfers from chair to bed, also Figure 85, as in any total pattern, head and neck lead and subject looks toward goal.

Preparatory mat activities include adversive rotation

A

B

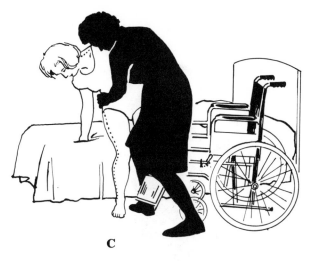

C

of upper trunk to right and lower trunk to left while lying on right side, balance side-lying (Fig. 44), sitting balance (Fig. 67), and standing balance (Figs. 73, 74).

Use of the developmental sequence

Fig. 87. From chair to standing to automobile.

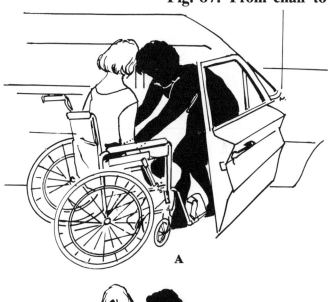

A

B

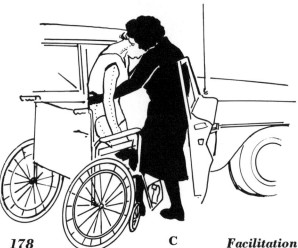

C

Component patterns

RESISTED

Lower trunk and lower extremities, extension with rotation toward right, then release to flexion toward right

FREE

Head and neck and upper trunk adjust toward flexion with rotation toward left

Left upper extremity thrusts in extension

Right upper extremity does not contribute to total pattern

A. Initial position

COMMANDS

"Lean forward, push with your left arm, and stand up!"

SUGGESTED TECHNIQUES

Approximation and resistance

B. Approaching shortened range, Standing

COMMANDS

"Hold it!" "Straighten your knees!"

SUGGESTED TECHNIQUES

Approximation

C. Approaching middle range, Sitting

COMMANDS

"Reach your left hand to the seat, and sit down slowly."

SUGGESTED TECHNIQUES

Guidance and resistance to lower trunk

Antagonistic pattern

Rising from sitting to standing to chair

Note: Subject is using extremities of left side to achieve transfer. Therapist has placed left knee against subject's left knee for stability of left lower extremity which is primarily responsible for control of inferior region.

Preparatory mat activities are as for transfer shown in Figure 86 with appropriate adjustments made for developing use and emphasis of right side of body.

SELF-CARE ACTIVITIES

By common definition, self-care activities are related to personal care, including feeding, using the toilet, bathing, and dressing. The normal child acquires these abilities through development and training. For the normal adult of sedentary occupation, bathing and dressing may constitute the most varied exercise of his daily life. The handicapped child needs to develop self-care abilities. The handicapped adult needs to relearn self-care. Because of their personal nature, self-care activities are of utmost importance to the patient's morale and motivation. In the total treatment of the patient, activities directed toward goals of self-care are given primary emphasis.

Training for self-care begins when the patient attempts to roll on a mat. The individuated patterns as used, for example, in feeding may have their origin in rolling from supine to prone and from prone to supine. Because the human being is capable of innumerable combinations of movement, each alteration of position places a different demand on the neuromuscular mechanism. To fully develop a self-care activity, performance of related component patterns in a variety of positions may be necessary to adequate performance of the activity.

With a good basis for function developed through mat activities and specific facilitation, a transition from gross to refined movement can be made. If, for example, a patient has achieved sitting balance and at the same time can move his upper extremities without disturbing his balance, it is obvious that he can more easily feed himself than if he were completely dependent upon a chair for support. If a patient can roll with ease and is able to use his extremities to accomplish the act, he can more readily learn to bathe and dress himself in bed.

The transition from gross to refined movements, such as those required for feeding, shaving, dressing the hair, and brushing the teeth, may be encouraged by superimposing techniques of facilitation on the functional movement or self-care activity itself. Performance in a position of functional use is desirable because it is conducive to eye-hand coordination. As coordination is acquired, the patient is free to attend other matters and his movements serve the matter which he is attending.

Resistance graded so as to encourage isotonic contraction of related muscle groups in the desired range of movement and repetition of a movement are adjuncts to training in self-care. Refinement of postural control for intricate movements is encouraged by performance of isometric contractions at specific ranges within a movement. Functional activities require reversing movements so that performance of reversal of antagonists techniques may hasten the learning of an activity. Other techniques, including relaxation techniques, may be used as necessary.

ILLUSTRATIONS

Figure 88, transfer from chair to toilet, and Figure 89, dressing in bed, are two examples of self-care activities. While all activities are related to self-care and independence, certain related in-bed activities (most of which are obvious) may be considered. The components of functional activities can be recognized more specifically if various combinations of elbow motions and reversal of direction of total patterns and of component patterns are visualized.

Turning, bathing, dressing	Rolling, Figures 38–48
Adjusting clothing at hips	Lower trunk activities, Figures 49, 50
Use of bedpan	Elevation of pelvis, Figure 50
Moving headward and footward	Crawling, Figures 51, 52
Reaching for feet	Sitting, Figures 65–67

Self-care is a goal of treatment. If the potential exists for performance of an activity in a coordinate manner, training in the position of functional use may be hastened by use of resistance and other techniques of facilitation. However, if imbalances exist, or if coordination is lacking, or if there is need to alter the rate of movement, these factors are not favorable to adequate self-care. The manifestation of bizarre and inadequate attempts to perform is evidence that the patient has been asked to perform beyond his level of ability and that he has need for additional work in more primitive patterns and in specific patterns of facilitation. Meeting this need may ultimately result in improved performance of self-care activities. The infant does not lace his shoes; he develops this ability through performance of a variety of less skilled activities.

SELF-CARE ACTIVITIES

Fig. 88. From chair to toilet.

A

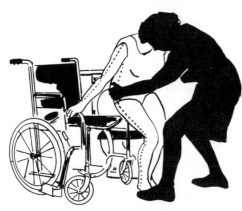

B

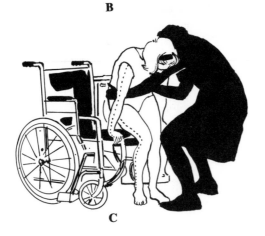

C

Component patterns

RESISTED

Lower trunk, elevation

FREE

Head and neck, and upper trunk adjust with flexion toward left

Upper extremities thrust downward, then alternately support and shift for progression toward left, then release with elbows flexing

Lower extremities do not contribute to total pattern

A. Initial position—Arm of chair removed

COMMANDS

"Reach your left hand to the toilet seat; now, push with your hands, and lift up."

SUGGESTED TECHNIQUES

Guidance and resistance for lower trunk

B. Middle range

COMMANDS

"Hold! Now, put your right hand on the seat of your chair."

SUGGESTED TECHNIQUES

Guidance and resistance, rhythmic stabilization

C. Approaching shortened range

COMMANDS

"Swing your hips to the left, and sit down slowly."

SUGGESTED TECHNIQUES

Guidance and resistance at pelvis

Antagonistic pattern

From toilet to chair

Note: Component patterns which may be trained with subject sitting in chair include: elevation of lower trunk, sideward rocking of pelvis while elevation is maintained, and reaching for seat of toilet. Where hand grips or rails are conveniently placed, training includes their use.

Transfer from chair to edge of bathtub or seat in shower stall may be carried out in a similar fashion including the use of hand rails. If a bench is to be used in a bath tub, transfers to benches of varying heights may be done as a preparatory mat activity.

As shown for transfer from chair to bed, Figure 86, subject may be trained for transfer to toilet. For transfer from toilet to chair, position of chair and of therapist must be adjusted.

SELF-CARE ACTIVITIES

Fig. 89. Dressing in bed—inferior region.

Component patterns

RESISTED

Reaching to wheel chair for slacks

Asymmetrical extension of upper extremities toward right lower extremity, left hand grasps ankle, right grasps slacks

Pulling slacks up to waist

FREE

Rolling to left with asymmetrical flexion of lower extremities

Flexion of head and neck, and upper trunk toward right

Extension of right lower extremity into leg of slacks

Rolling to right with head and neck rotation

A. Reaching to chair

COMMANDS

"Reach, take hold of your slacks, and pull them to you."

SUGGESTED TECHNIQUES

Resistance, slow reversal

B. Placing right lower extremity

COMMANDS

"Take hold of your right ankle with your left hand, and your slacks with your right. Pull your right foot toward you some more." "Don't let me pull your right hand off your slacks." "Now, push your right leg down with your left hand, and pull your slacks up with your right."

SUGGESTED TECHNIQUES

Resistance during various phases

C. Adjusting slacks at waist

COMMANDS

"Pull with your left hand." "Now, roll toward me, then pull with your right."

SUGGESTED TECHNIQUES

Resistance during various phases

Antagonistic pattern

Undressing inferior region

Note: Component patterns to be trained include: asymmetrical flexion of lower extremities with rolling to left, thrusting of right upper extremity from shoulder extension with elbow flexion, *A*; "chopping" to right

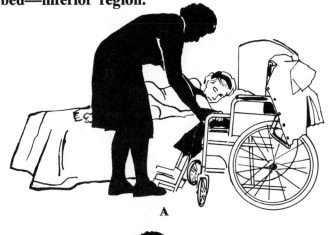

A

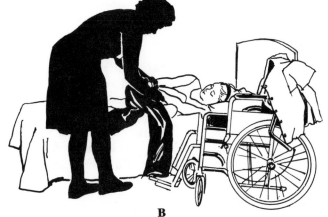

B

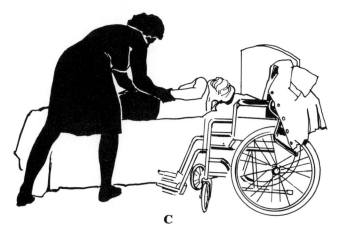

C

with elbow extension, *B*; rolling to right and to left, and, if possible, lower trunk activities, elevation of pelvis and lower trunk rotation, Figures 49 and 50.

Manipulation of zippers, buttons, and Velcro fastenings may be resisted. Also, of course, putting on and removal of socks and shoes, and braces may be trained by use of resistance. Sitting over edge of bed, or in chair may be assumed for dressing of superior region.

SUPPLEMENTARY CLASS ACTIVITIES

In the process of development and in motor learning, the child practices and uses the total and component patterns he has mastered. He may try and fail but he repeats his efforts until he succeeds. Once he has learned an activity, it is a part of him. He can use it automatically or deliberately, as the occasion demands. The handicapped child or adult needs opportunity to practice the activities he is learning and to use those he has mastered. In this way he advances his own progress.

Supervised classes provide an opportunity for continued motor learning, for establishing more firmly those patterns of movement which have been learned, and for developing strength, endurance, and stability of posture. Individual patients have individual needs. The activities they perform in a class are selected to meet their needs and to fulfill specific goals. Several patients may have the same needs so that they may work together or compete with each other, but the concept of class work is supervision of individual programs of activity rather than group work of selected activities. Patients perform in different ways and at different speeds.

Class work carried out on gymnasium mats may include free activity and mechanically resisted activity. Free activity is based on the developmental activities. When possible, gymnasium equipment, including dumbbells, medicine balls, beach balls, and pulley weights are used for increasing demands and challenging balance or stability of posture (ref. 28). The patient is limited to those activities which he can do in good form in accordance with normal timing; he is limited to coordinate movements or those movements which will improve his coordina-

tion. The supervisor is a teacher, a taskmaster, and sometimes an arbitrator. Patients are people. They play together, they compete, they win, they lose, they challenge each other, they help each other (ref. 14). They learn again to be members of a group, to overcome self-centered attitudes. The wise supervisor enlists those patients who can help others. The hemiplegic adult may practice hand-knee balance while talking to and stimulating a brain-injured child. An active child competes with adults and challenges them. They may play together and may be taught to resist each other in useful movements. In a sense, a mat class is not a class but a community of activity.

Class work carried out by use of gymnasium wall pulleys, too, is limited to those patterns of movement which can be performed in a coordinate manner. The specific patterns of facilitation and combinations of patterns are used. In some instances, the weights must provide assistance to a movement. In other instances, resistance of the antagonistic pattern may encourage greater range of a desired movement. Patients may perform in various positions on a treatment table, or sitting in a chair, or standing. The individual patient is taught and asked to perform those movements which support the goals of his treatment. The supervisor, or teacher, must direct activity, determine the amount of weight to be used, and help patients to alter positions and activities as necessary. Again, the supervisor is a taskmaster. Some patients respond to coaxing; others respond best to the firm commands of an overseer. Interplay through communication with other patients has potential benefits. The wise supervisor assigns patients to locations of mutual benefit.

4. Stimulation of vital and related functions

Stimulation of vital and related functions

Vital and related proximal functions may be defined as those functions of the body which are primarily under reflex control but which may be inhibited at will. They include motions of respiration, facial motions, eye motions, opening and closing of the mouth, tongue motions, swallowing, micturition, and defecation. Performance of patterns of facilitation against maximal resistance stimulates related motions which have to do with proximal functions.

Beyond the stimulation achieved during performance of related patterns, the techniques of proprioceptive neuromuscular facilitation may be specifically applied to the motions of the parts responsible or necessary to the vital functions. As with all movement, these functions may and should be stimulated in a variety of positions. For example, breathing motions may be resisted in prone and side-lying (lateral) positions as well as in supine. Swallowing is easier in the prone position than in supine. Tongue motions may be encouraged more effectively when the patient is prone with chest elevated by support on elbows and forearms. Where deficiency is marked, the most favorable position should be sought.

Analysis of individual muscles is not included but study of these muscles will reveal that they are, in general, aligned in a spiral and diagonal fashion.

RESPIRATION

Techniques of proprioceptive neuromuscular facilitation may be applied as a means of stimulating response and strengthening muscles related to respiration. Strengthening of neck, trunk, and extremity patterns has a by-product of increased ability in respiration. The patterns most closely related to inspiration are neck extension, upper and lower trunk extension, and flexion patterns of the upper extremities. The patterns most closesly related to expiration are neck flexion, upper and lower trunk flexion, and extension patterns of the upper extremities.

Combinations of these patterns such as upper trunk motions combined with bilateral asymmetrical upper extremity patterns (chopping and lifting), and bilateral symmetrical upper extremity patterns simulate stress situations. An increased demand is placed upon the accessory muscles of respiration which are normally used in deep respiration as well as the respiratory mechanism itself.

Stimulation of the intrinsic muscles of respiration and increased range of motion of the chest and diaghragm are achieved by direct application of techniques of facilitation. Resistance may be applied to the motions of the lateral chest walls, upper chest, sternum, and diaphragm. Correction of imbalances is approached by maximal resistance to a stronger area and repeated contractions emphasizing the weaker area.

For example, if a patient has stronger response of the left lateral chest wall than of the right lateral chest wall, resistance is applied as follows: The physical therapist places his hands on the lateral chest walls with the bases of the palms near the apices of the lower ribs and with the fingers lying in close approximation in an upward and lateralward direction. The physical therapist instructs the patient to exhale, "breathe out," and applies pressure downward and medialward so as to achieve stretch on the intercostal muscles. The patient is then instructed to "breathe in," "as much as you can," "hold it!" As the patient inhales, the physical therapist lessens the pressure and grades resistance so as to encourage range of motion. When the patient "holds" his breath, the physical therapist proceeds to repeated contractions. While the therapist maintains steady resistance to the stronger chest area he repeatedly alternates increasing and decreasing pressure over the weaker area. The patient is instructed to "breathe in again, and again, and again" and attempts to sustain his effort throughout the procedure. When the patient has repeated as many times as he is able, he exhales in a sustained manner.

The above procedure may be applied to emphasize the upper region of the chest by using one hand in contact with the sternum and the other hand and arm over the lower lateral chest walls. The hand which is used for sternal contact is placed with the base of the palm on the manubrium sterni, the fingers are placed in close approximation to each other extending downward toward the xiphoid process. Pressure is applied in a diagonal direction—downward and toward the abdomen. Pressure should not cause pain and if the patient experiences pain, the pressure has been too great in a directly downward direction. The other hand and arm are used to compress the lower chest, thereby channeling air to the upper region of the chest. Repeated contractions may be performed with the hand which is in sternal contact.

Emphasis of either side of the upper region of the chest may be performed by placing both hands with the bases of the palms near the sternum and the fingers pointing upward and outward toward the acromion processes. Various combinations of one area of the upper region of the chest and an area of the region of the lower chest may be used. The stronger area is used to reinforce the weaker area. This is done by preventing motion in the stronger area by pressure and resisting the weaker area, grading resistance through the range.

Stimulation of the diaphragm is accomplished by placing the thumbs and palms of the hands along the costal cartilages of the lower ribs. Pressure and stretch is applied with the thumbs pushed up and under the rib cage as far as possible without producing pain. The tips of the thumbs are pointed toward the xiphoid process. Repeated contractions may be performed to both sides simultaneously or one side may be emphasized with sustained pressure to the other side. Resistance may be applied to forced expiration in this area by resisting the downward motion of the rib cage so as to prevent the patient from decreasing the diameter of his lower chest as he exhales.

Rhythmic stabilization may be performed as stimulation for the diaphragm by using the thumbs in contact as described above. The fingers are placed in contact with the lower chest walls. The patient is instructed to "breathe in," "hold it!" The patient sustains his breath while the physical therapist applies pressure and stretch alternately to the chest walls and the diaphragm. After two or three alternations, the patient is instructed to "breathe in again! again! and again!" while the physical therapist repeats with increasing and decreasing pressure to the diaphragmatic area.

Success of application of techniques to patterns of respiration will depend upon the physical therapist feeling the patient's response, synchronizing his demands with the patient's efforts, and carefully grading resistance so as to encourage response and range of motion.

FACIAL MOTIONS

The normal facial motions are bilaterally symmetrical in character—both sides of the face move in identical motions. The normal subject is capable of innumerable combinations of facial motions which include unilateral and bilaterally asymmetrical motions. While the normal subject may isolate certain motions to a degree, in situations of emotional stress facial motions are usually bilaterally symmetrical in character. During vigorous physical activity facial motions may take on a bilaterally asymmetrical character in that facial motions are brought into play as reinforcement. Inability to perform bilaterally symmetrical motions voluntarily is an indication of weakness resulting in asymmetry of facial expression.

Facial motions may be grouped as antagonistic motions involving three pivots of action—the mouth, the nose, and the eyes. Extreme ranges of motion of any one pivot brings into play related movements of other pivots. Antagonistic motions may be considered as follows:

1. Elevation of eyebrows, upward and lateralward —Depression of eyebrows, downward and medialward
2. Opening of eyelids, lateralward—Closing of eyelids, medialward
3. Elevation and opening of nostrils, lateralward —Depression and closing of nostrils, medialward
4. Retraction of angles of mouth, upward—Protrusion of lips, downward
5. Retraction of angles of mouth, downward—Protrusion of lips, upward
6. Opening of lips with protrusion—Closing of lips with inversion

The facial muscles are spiral and diagonal in character and are arranged for symmetrical motions. Strong contraction of the circular muscles about the mouth and the eyes demands lengthening or shortening reactions of the other facial groups including those of the scalp. Strong contraction of the nasal groups in turn demands cooperation of the muscles responsible for motion about the eyes and the mouth.

The various techniques of proprioceptive neuromuscular facilitation that may be applied to facial motions include pressure, stretch, resistance, reinforcement, repeated contractions, and reversal of antagonists. Relaxation techniques may be used as indicated. The physical therapist uses the tips of the fingers as manual contacts. Stronger motions are resisted in order to stimulate and reinforce weaker motions.

For example, a patient may present weakness of elevation of the left eyebrow. The physical therapist places her finger tips on both sides of the patient's brow and applies pressure and stretch in a downward and medialward direction. Having achieved stretch she instructs the patient to "Look up at me! Raise your eyebrows!" At this point the physical therapist resists strongly the motion on the right and allows range of motion to occur on the left. The patient is instructed to "hold" his eyebrows raised, and the physical therapist then applies the technique of repeated contractions. Reversing motions and relaxation techniques may be used to increase mobility of the elevation and depression of the brows.

Motions of the eyebrows may be used to reinforce opening and closing of the eyes, motions of the lips may be used to reinforce motions about the nose or the eyes. Study of the normal subject will reveal the relationship of facial motions. Neck patterns may be used as reinforcement. Any facial motion which requires elevation or upward motion is reinforced by neck extension. Facial motions which require depression or downward motions are reinforced by neck flexion. Neck rotation reinforces the motion of the side of the face to which the head is turned. If it is desired to reinforce a motion on the left, the head is turned to the left.

EYE MOTIONS

Performance of related neck and upper extremity patterns where the eye follows the hand provide stimulation for movement of the eyes. Neck extension with rotation patterns reinforce upward and lateralward eye movements. Neck flexion with rotation patterns reinforce downward and lateralward eye movements. Neck rotation reinforces lateral movements. The lateral movement of a given pattern determines the lateral direction of the eye movement—to the left or to the right.

Eye movements may be stimulated, certain movements or ranges of movement may be emphasized by application of techniques of facilitation. The physical therapist may use her index finger or a pencil and instruct the patient to follow the movement of the object. Motions may be upward, downward, or lateralward and in various combinations, such as upward and lateralward to the left or right, or downward and lateralward to the left or right. Reversing movements may be applied or repeated contractions may be performed. In order to perform repeated contractions, the physical therapist allows the patient to perform in one direction as far as possible. At this point, the physical therapist moves the pencil in the opposite direction very slightly and just as the patient is about to follow with his eyes, the physical therapist quickly moves the pencil in the original direction. Related neck patterns may be performed actively or against resistance as reinforcement.

OPENING AND CLOSING OF THE MOUTH

Opening of the mouth requires depression with retraction of the mandible, closing of the mouth requires elevation with protrusion of the mandible. The mouth may be opened or closed in mid-line, or it may be opened and closed to one side or the other side, combining lateral movements with opening and closing.

In patterns of facilitation, opening is related to neck flexion patterns and closing is related to neck extension. Lateral motion of the mandible is related to neck rotation. When opening of the mouth is reinforced by neck flexion with rotation to the right, depression with retraction and lateral motion of the mandible toward the right occurs. When closing of the mouth is reinforced by neck extension to the left, elevation with protrusion and lateral motion of the mandible toward the left occurs. When lateral motion of the mandible is reinforced by neck rotation, a certain amount of mandibular depression occurs. If reinforcement of mandibular motions by neck motions is attempted, the head should be free of hard surface contact and the manual contacts described for the neck patterns may be used.

Techniques which may be used include pressure through manual contacts, stretch, resistance, reinforcement, repeated contractions, and reversal of antagonists. If range of motion is limited by adaptive shortening or contracture, relaxation techniques may be applied.

MOTIONS OF THE TONGUE

The tongue is an extremely versatile part of the body when its repertoire of motions and dexterity is considered. Elevation, depression, protrusion, retraction, lateral motions, and rotatory motions are combined in various tongue motions. Combining several motions during evaluation will often disclose asymmetry of function or imbalance between the two sides. Protrusion straight forward may be combined with elevation and depression. Retraction straight backward may be combined with elevation and depression. Protrusion and elevation laterally to one side may be compared with protrusion and elevation laterally to the opposite side. Protrusion and depression laterally to one side may be compared with the same motion to the opposite side. In the same manner, retraction may be combined with depression, elevation, and lateral motions.

Resisted neck motions encourage reinforcement of those motions by motion of the tongue. Neck extension in turn reinforces elevation of the tongue, neck flexion reinforces depression, and neck rotation reinforces lateral motion. Opening of the mouth is delated to depression of the tongue and closing is related to elevation.

Techniques of facilitation may be applied to tongue movements for the purpose of strengthening the movements of the tongue and correcting imbalances. The patient's tongue may be grasped by the physical therapist's gloved fingers or a piece of gauze. A tongue depressor blade may be used in resisting certain motions but this method is less satisfactory since the motions cannot be as well controlled.

For example, if protrusion and elevation laterally to the left is weaker than that combination of motions to the right, the physical therapist places her fingers so as to push the patient's tongue backward, downward, and to the right. The patient is instructed to pull his tongue backward as the physical therapist places her fingers. After the tongue has been pushed backward, the physical therapist instructs the patient to "Push your tongue out and up to the left, and hold it there!" The physical therapist resists the motion and may perform repeated contractions as the patient sustains his efforts, or reversing motions may be performed. The physical therapist may resist related neck motions or opening and closing of the mouth as reinforcement, but caution must be used to prevent the patient from biting his tongue or the fingers of the physical therapist.

SWALLOWING

Swallowing or deglutition is a complex act requiring interaction of the suprahyoid and infrahyoid muscle groups. These same muscle groups augment the neck flexion patterns and receive stimu-lation when these patterns are performed. However, when neck patterns are performed against resistance, it is almost impossible for the normal subject to perform swallowing movements simultaneously. When the head is positioned in the lengthened range of either neck flexion or neck extension, swallowing becomes difficult.

Resistance may be applied to swallowing motions through the use of a simple device. A piece of sponge rubber, approximately $\frac{1}{2}'' \times \frac{1}{2}'' \times \frac{1}{2}''$, may be tied securely to a piece of stout cord or string. The rubber is placed upon the patient's tongue and he is instructed to "swallow" while the physical therapist provides resistance by tugging at the cord attached to the rubber.

Stimulation of elevation of the soft palate may be accomplished in the following manner. The patient opens his mouth widely and says "Ah." The physical therapist touches the uvula or soft palate folds on either side with a cotton-tipped applicator stick which stimulates a reflex contraction. The patient repeats "Ah" as many times as possible. As a lag of response appears in an area, the physical therapist again stimulates the area by touching it lightly and as often as necessary.

Stimulation of the gag reflex is helpful in that it demands response of the pharyngeal muscles with elevation of the soft palate and is followed automatically by a swallowing response.

MICTURITION AND DEFECATION

The voluntary performance and control of micturition and defecation may be enhanced by performance of related patterns of facilitation against maximal resistance. The act of emptying the bladder or bowel is most closely related to the flexion patterns of the lower trunk and lower extremities. Inhibition of these acts is most closely related to extension patterns of the lower trunk and lower extremities.

The muscles of the perineal region may be stimulated during the performance of bilaterally symmetrical extension–adduction–external rotation patterns of the lower extremities. Specific stimulation of these muscles may be obtained through application of stretch and resistance. The patient should lie supine with the lower extremities flexed in abduction. If the patient is able, he may maintain his own extremities in the described position and may himself resist the extension–adduction–external rotation of his lower extremities. If he is not able to use his hands for the purpose of supporting and resisting his extremities, his feet may be supported on the table. The physical therapist may then apply stretch to the perineal region. Stretch is given in an upward and outward direction from the anus. The physical therapist instructs the patient to pull his extremities down and together. To stimulate muscles of the perineum, the extremely lenghtened range of extension–adduction–external rotation patterns must be used. Preventing all movement except external rotation will achieve the desired effect. The physical therapist resists the contraction of the muscles and repeated contractions may be performed while the patient sustains his effort.

5. Evaluation and treatment program

Evaluation of patient performance

The evaluation of patient performance must be based upon knowledge of performance of normal subjects. While normal subjects vary with regard to available range of motion, coordination, power, endurance, and rate of movement, the variations are within normal limits. These variations do not affect the subject's adequacy for ordinary motor activities, but may affect the subject's ability to perform highly skilled activities, his postural attitudes, his power, and his endurance.

The mature, normal subject is able, after verbal instruction, to:

1. Initiate all patterns of facilitation from the lengthened range and to proceed to the shortened range (isotonic contraction) in accordance with normal timing.

2. Perform all patterns against maximal resistance in accordance with normal timing.

3. "Hold" at any desired point in the range of motion (isometric contraction). In the shortened range he is able to "hold" so strongly that the "hold" cannot be broken unless derotation is used.

4. Perform all combinations of related patterns.

5. Perform reversals of patterns and the various techniques of facilitation against maximal resistance with a resultant build-up in power or gain in range of motion.

The physical therapist may defeat the performance of the subject by preventing rotation from occurring in the lengthened range of the patterns; by preventing normal timing through application of excessive resistance; or by derotating the part when a "hold" contraction is being performed.

It must be remembered that physical therapists as a group of normal individuals present variations of performance as clearly as another group of normal subjects. While serving as a subject, the physical therapist must accept a goal of performance rather than a goal of self-analysis. It the subject directs his attention to such things as which muscles are contracting, he hinders his own performance and frustrates his physical therapist. Analysis and critical evaluation are helpful after performance.

GENERAL OBJECTIVES

The goal of evaluation or analysis of the patient is to ascertain his abilities, deficiencies, and potentialities accurately. Necessary general knowledge includes:

1. Considering the patient's chronological age, are his motor abilities adequate?
    a. If they are inadequate, are the deficiencies the result of:
    faulty development;
    trauma;
    disease?
    b. Are deficiencies evidenced by:
    less or more than normal range of passive and active motion;
    incoordination;
    less than nomal strength;
    lack of endurance;
    less or more than normal rate of movement;
    instability of posture?
2. Is deficiency generalized and profound or is deficiency most evident in relation to the proximal parts (neck and trunk) or the distal parts (extremities)?

PATTERNS OF FACILITATION

Detailed evaluation of the specific patterns of facilitation is most easily done with the patient lying on a treatment table. However, those patterns which are not limited as to range when performed on a flat surface may be evaluated as the patient lies on a gymnasium mat or as he performs total patterns of the developmental sequence.

Specific information to be gained by evaluation includes:

1. Which patterns of facilitation are within normal limits? Which patterns are inadequate?

2. Is the available range of motion of specific patterns within normal limits?
   a. Is passive range limited by:
      adaptive shortening or contracture;
      spasticity;
      muscle spasm;
      pain?
   b. Is passive range of motion excessive?
      Are the muscles lengthened beyond their normal limits so that they do not serve as a range-limiting factor to the antagonistic pattern?
      Have the ligaments, joint structures, and soft tissue contact become the only range-limiting factors?
3. Is active performance in keeping with the developmental level of the subject, considering his chronological age?
   a. Is performance smooth or is motion of the distal parts delayed?
   b. Is performance in the "groove" of the pattern or do bizarre movements occur?
4. Is performance of isotonic contraction possible through the available range of motion?
   a. Is the active range less than the available range of passive motion?
   b. Is active range limited by adaptive shortening or contracture, spasticity, muscle spasm, pain?
   c. Are the major muscle components of less than normal strength?
      Is deficiency present throughout the pattern or is it most evident with relation to proximal, intermediate, or distal pivots?
      Is a specific component of action more deficient than the other components, i.e., flexion, adduction, or external rotation of the hip in the flexion–adduction–external rotation pattern?
5. Is the performance of isometric contractions deficient?
6. Is a pattern more or less deficient than its directly antagonistic pattern—i.e., flexion–adduction–external rotation pattern of the lower extremity as compared with extension–abduction–internal rotation pattern of the same extremity?
   a. Is imbalance most evident at proximal, intermediate, or distal pivots of action?
7. Is a pattern more or less deficient than its related pattern of the opposite diagonal—i.e., flexion–adduction–external rotation pattern as compared with flexion–abduction–internal rotation pattern of the same extremity? Proximal, intermediate, or distal pivots?
8. Is a pattern more or less deficient than the antagonistic pattern of the opposite diagonal, i.e., flexion–adduction–external rotation pattern as compared with extension–adduction–external rotation pattern? Proximal, intermediate, or distal pivots?

Plan of Approach

In order that abilities, deficiences, and potentialities may be accurately disclosed, a systematic plan of approach is used. The following plan is general in nature and must be adapted for the individual patient. Contraindications for attempted full range of passive motion, active motion, or motion against resistance will necessarily alter the approach to evaluation.

The sequence of evaluation is from proximal to distal. The proximal parts and function of these parts deserve first consideration since they are related to vital functions of the body. Whenever there is a possibility for deficiency of proximal function related to respiration, tongue motions, soft palate reactions, swallowing, and facial motions, these are evaluated first. Since the neck patterns are the key to the upper trunk, they receive the next consideration. The sequence of evaluation then proceeds as follows: upper trunk, upper extremities—proximal, intermediate, and distal pivots, lower trunk, lower extremities—proximal, intermediate, and distal pivots.

Accurate evaluation requires considerable time and effort on the part of both the patient and the physical therapist. It may be necessary to devote several sessions to evaluation because of the time factor or the fatigue factor (physical therapists frequently tire before a patient will admit fatigue). The first session may be limited to evaluation of proximal parts and upper extremities. The second session may then be devoted to a brief review of the proximal parts and evaluation of the lower extremities.

PASSIVE MOVEMENT

A specific pattern should first be checked for available range of passive motion. Limitation of range or excessive range should be noted for each pivot of motion and the part of the range where the variation becomes evident. The extremity patterns which demand complete lengthening of two-joint muscles should be compared with the pattern which does not require lengthening of these muscles. This comparison is necessary in order to determine the pivot of motion which is most limited. For example, evaluation of passive range of the flexion–

adduction–external rotation pattern with reference to the hip pivot demands that the part be moved with the knee straight and with the knee flexed. Unless this is done, normal length of the biceps femoris may mask a limitation in range of motion at the hip.

Passive movement of the part is performed in distal to proximal sequence. The distal pivots are moved into their shortened range, then intermediate pivots, then proximal pivots. Increased tension occurring at distal and intermediate pivots as the shortened range of the proximal pivot is approached should be noted. Such increase in tension may indicate intermediate and distal limitation. Movement may be repeated without demanding full range of the distal pivots in order to determine the degree to which distal tension limits proximal range of motion. This overlapping of tension is due to the topographical interrelationship of the major muscle components. Marked limitation of passive range of one pivot may influence passive range of the other pivots of motion. Excessive range of passive motion or hypermobility of a specific pivot of motion must also be noted and given due consideration in the treatment program.

ACTIVE MOVEMENT

The range of active motion of a specific pattern may be considered immediately following the evaluation of passive range. The patient should be instructed and asked to perform with all combinations of intermediate joint motion. Performance should be observed for variations in timing and range of active motion of each pivot of action. Comparison of active range and passive range should be made. Repetitions of performance may verify deficiencies or may reveal a lack of understanding by the patient of the physical therapist's instructions. Performance of a pattern against gravity may be considered as a basis for evaluation of active motion, and the patient may be positioned accordingly. Positioning must allow for the full range of motion to occur at the various pivots.

RESISTED MOVEMENT

Unless contraindicated, maximal resistance may next be superimposed as a means of determining specific deficiencies. The patient should be required to perform with isotonic contraction through full range or as much range as possible, in accordance with normal timing, and should be required to perform an isometric contraction in the shortened range of the pattern. Resistance must be graded so

as to permit normal timing to occur. Variations in timing, the range of motion of the specific pivots, and the strength of isotonic contraction and isometric contraction should be observed.

FACILITATED RESPONSE

If the patient is unable to perform the full range of the pattern, or if a specific pivot fails to perform adequately, timing for emphasis and stretch stimulus may be used to determine potentials for increased response of the weaker pivots of action. The response of a weaker pivot during timing for emphasis should be compared with the response felt at this pivot when normal timing was performed against resistance. Increased response of a weaker pivot during timing for emphasis indicates potential for improvement, since overflow or reinforcement is apparently evident. This pivot may then be considered as a pivot for emphasis during the treatment program. If there is no increase in response of the weaker pivot during timing for emphasis, deficiency is great and emphasis of this pivot must be delayed until proximal parts, or the more proximal pivots, have been strengthened sufficiently to provide reinforcement for the weaker pivot.

DEVELOPMENTAL ACTIVITIES

Inasmuch as developmental activities are total patterns of movement with interaction of body segments, evaluation of the patient's abilities, disabilities, and potentials for performance is necessarily done on a gymnasium mat or in an open area. These activities demand space so that repetition of total patterns may be observed. In evaluation of the specific patterns of facilitation, interest is centered primarily upon one segment at a time. In evaluation of activities in the developmental sequence it is necessary to observe the total structure in a total pattern of movement. Furthermore, it is necessary to observe the coordination of component patterns. Having assessed the performance of patterns of facilitation, a certain understanding of the patient's abilities and disabilities has been gained. This knowledge is useful in observation and assessment of a total pattern of movement.

In general, the objectives and the questions posed in evaluation of patterns of facilitation apply in evaluation of developmental activities. When the component patterns of a total pattern are individually assessed as patterns of facilitation, the information to be gained is the same. Yet, the total pattern must be analyzed as a total pattern. The following questions should be raised in respect to the three areas of use—mat activities, gait activities, and self-care activities.

1. Is the performance of a total pattern or activity in keeping with the developmental level of the subject, considering his chronological age?
2. If performance is inadequate, what are the major problems and areas of deficiency? Is performance limited by:
   a. inability to respond to verbal commands or appropriate auditory stimuli, visual cues, or proprioceptive stimuli?
   b. generalized weakness, incoordination, spasticity, spasm, rigidity, pain, or multiple contractures?
   c. inadequacy of component patterns of head and neck, upper trunk and upper extremities, or lower trunk and lower extremities?
   d. inadequacy of ipsilateral component patterns, or of bilateral component patterns of upper or of lower extremities?
   e. inadequacy of distal parts as compared with proximal parts?
3. Are those activities having flexor dominance more adequately performed than those activities having extensor dominance?
4. Is the maintenance of a balanced posture adequate or inadequate as compared with the ability to assume the position of balanced posture? Is the ability to move greater than the ability to sustain posture?
5. Considering the patient's ability to perform certain specific patterns of facilitation on the treatment table, is his ability to perform the same patterns during mat and gait activities more or less adequate?
6. Considering the patient's inability to perform certain specific patterns of facilitation on the treatment table, is he able or unable to perform the same patterns during mat and gait activities? Does alteration of position permit him to perform a certain pattern; that is, if he is unable to perform a certain pattern in the supine position, is he able or unable to perform in prone, lateral or side-lying, sitting, or standing positions?
7. Considering the patient's ability to perform a certain specific pattern of facilitation on a treatment table or during a mat activity, is he able to perform in a position of functional use? Is he able to combine and perform related component patterns necessary to a self-care activity?
8. Does the patient adequately reinforce his attempts to perform a total pattern of movement? Would external support for weak or inadequate segments permit him to better reinforce his attempts to move or to sustain posture?

Plan of Approach

The progression of activities as outlined in the developmental sequence, Table 2, serves as a guide for evaluation. Those activities which can best be performed on a mat should be evaluated as a mat activity. Those related to gait may be performed on a mat, if possible, but most often severely disabled persons require the use of other equipment—wheel chairs, parallel bars, crutches, canes, braces, or assistive devices. Self-care activities should be assessed with the patient in a location appropriate to the activity: in bed; in his wheel chair, if he uses one; and in the bathroom. Special areas designated for practice of gait and self-care activities may be used for assessment but the patient's specific problems may become more evident in the area where he actually lives, whether in hospital or at home.

Just as there is an overlapping between activities within the developmental sequence so is there an overlapping of information gained by evaluation of various activities. The evaluator will learn which component patterns can be performed by the patient and those he cannot perform. The evaluator will interpret the patient's performance in relation to mat, gait, and self-care activities. The evaluator will sort and classify information about the patient. The evaluator will identify the patient's abilities and his disabilities. The evaluator will seek realistic goals according to the patient's abilities and his potentials for increasing his abilities. The evaluator will develop a plan of treatment for the patient or will communicate as accurately as possible with those persons responsible for the planning and execution of the treatment program.

Planning the treatment program

The diagnosis, indications for treatment, and goals of treatment are established by the member of the medical profession who is responsible for the patient. Exercise programs are planned with regard for the diagnosis, indications, and contraindications for exercise. Goals or objectives of treatment as established by the physician are usually general in nature. The over-all objective of an exercise program is to hasten recovery of normal function, to attempt to establish or reestablish optimum function as quickly as possible. Beyond these facts, specific areas and pivots for emphasis must be established, and techniques must be selected as a means of helping to achieve or surpass the goals as outlined by the physician. The patient may be treated on a mat rather than on a table since resistance, repeated contractions, and reversal techniques may then be superimposed upon total patterns of movement of the developmental sequence. In this way initial power may be developed and stimulation of mass patterns of flexion and extension may be achieved. A therapeutic pool provides an excellent opportunity for treatment of selected adult patients, since combinations of trunk motions may be performed with the body weight free of hard surface contact. The physical therapist is able to handle the adult patient as if he were a small child. Thus the patient may be treated in whatever environment is appropriate for his condition and his needs.

AREAS FOR EMPHASIS

Selection of areas of emphasis and specific pivots of emphasis is based upon the evaluation of the patient, which provided an impression of the patient's abilities, deficiencies, and potentialities. Emphasis in treatment means that more time and effort is directed toward certain areas of the body or pivots of action than toward other areas or pivots of action. Shifting of emphasis from one pattern to another pattern and from one area of the body to another area serves the first area as recuperative motion and the second area as stimulation of neuromuscular response. In line with the process of development, selection of areas and pivots for emphasis is from proximal to distal. Any deficiency of proximal parts receives the first emphasis. As these parts improve, they provide reinforcement for the extremities. The proximal pivots of the extremities receive next emphasis, and then the more distal pivots. Available power in the distal parts may be used to reinforce weaker proximal pivots, but the emphasis is still proximal. If generalized weakness exists, the sequence from proximal to distal is mandatory.

If the neck, trunk, and extremities all present marked deficiency, the treatment program will include all combinations of neck and upper trunk patterns, upper and lower trunk patterns. Consideration should be given to correction of imbalances, but the stronger patterns must be used as reinforcement of the weaker patterns. If a patient presents trunk deficiencies but has some available power in the extremities, the extremity patterns which are most closely related to the trunk patterns may be used as reinforcement.

If a patient presents greater strength in the lower trunk and lower extremities than he does in the upper trunk, neck, and upper extremities, the lower extremities and lower trunk are used to reinforce the upper trunk and neck. The upper trunk and neck are the areas for emphasis until such time that they will reinforce the upper extremities in their related patterns. The stronger lower extremities may be used to reinforce the upper extremities if such reinforcement enhances the response of the upper extremity pivots. It is frequently possible to achieve greater response of the neck and upper trunk by reinforcement of the lower extremities, if the patient is treated in a standing position. The patient automatically reinforces resisted motions of the neck and upper trunk by employing postural and righting reflexes. Rhythmic stabilization, reversal techniques, and repeated contractions for emphasis may be used with resistance applied to the head and neck.

A patient whose deficiencies are confined to one upper extremity has many potentials for reinforcement. Emphasis on the scapular motions or the proximal pivot is mandatory since scapular stability is essential to total function of the shoulder. The neck and opposite upper extremities provide the ideal reinforcements. The patient may be treated in any position which allows for performance of the desired patterns.

The above examples are given as a means of emphasizing the importance of proximal stimulation and the use of stronger areas as reinforcement for weaker areas. Areas for emphasis do not preclude over-all stimulation. When large combinations of patterns are performed, mass stimulation occurs through irradiation and overflow. An extremity which presents response, however minimal, deserves stimulation, but that extremity cannot be considered as an area for emphasis if the trunk is also grossly deficient.

PIVOTS FOR EMPHASIS

The selection of specific pivots for emphasis is also proximal to distal. In the extremities, the shoulder girdle and hip receive first emphasis, then the intermediate and distal pivots. A balance of power between antagonistic patterns is of paramount importance, and imbalances are also corrected from proximal to distal. If the scapular motion is weak in the extension–abduction–internal rotation pattern and all other scapular motions are strong, the weak scapular motion is the first pivot for emphasis to be considered in the entire extremity. It is futile to emphasize weak opposition of the thumb unless any deficiency of internal rotation of the shoulder has been corrected. Opposition of the thumb is primarily a motion of rotation, and reinforcement for this distal rotation is dependent upon intrinsic internal rotation of the shoulder.

SELECTION OF TECHNIQUES

Selection of techniques cannot be made on an arbitrary basis except in relation to contraindications for a specific technique. Application of stretch stimulus is obviously contraindicated in the early treatment of fractures and post-operative conditions. Decision as to a choice of techniques must be based upon the patient's response to that technique. The technique which facilitates a desired response to the greatest degree is the technique of choice. In general, if a patient needs to develop his ability to move, techniques that use isotonic contraction of muscle are employed; if he needs to develop postural control, techniques for isometric contraction are used. Coordinated performance of reversal of antagonists, through the full range of the patterns with normal power, is a response of the normal subject. Use of this technique should be a goal if it cannot be used initially. Decision as to reinforcements, either of one pivot of action by another pivot of action, or one pattern by another pattern, must be based upon the relationship of the pattern, the available power, and the response of the reinforced pivot or pattern. Reinforcement may be limited to the use of stronger pivots of action within the weaker pattern of a single extremity, if imbalances within the pattern are so great that control of the motion requires both hands of the physical therapist. As soon as control of the part is possible, larger combination of motion may be used.

That patient progress may proceed at an optimum rate the physical therapist should consider:

1. Proximal deficiencies should be corrected first, since proximal power provides more effective reinforcement.

2. Development of a balance of power in relation to all patterns and all pivots of action is of paramount importance. A balance of power implies adequate performance of isotonic and isometric contractions of antagonistic patterns.

3. Reinforcements should be selected in accordance with the increase in response they provide. Correction and prevention of imbalances should be considered, but stronger motions must be used to stimulate weaker motions. Wise selection of pivots for emphasis and reinforcement will prevent increase of imbalances. Selection of reinforcements may be influenced by the physical therapist's ability to control the combination of motions.

4. The choice of a technique should be determined by the patient's response. Arbitrary selection is undesirable except as indicated by contraindications.

5. Frequent re-evaluation of the patient is essential, so that new areas of emphasis may be included and unnecessary ones eliminated.

INTEGRATION OF ACTIVITIES

For optimum development or restoration of a patient's neuromuscular abilities, coordination of various phases of treatment is necessary. All activities must be integrated and directed toward the goals set for the individual patient. Goals may be expressed in terms of gait or self-care abilities, but the means to their accomplishment may be treatment on a table or on a mat or in performance of mat class and pulley class activities.

Carefully selected activities are performed in the location which permit the patient to put forth maximum effort so as to learn at an optimum rate. Economy of the therapist's effort and time are factors. Performance of developmental activities on a mat may contribute more in less time than will performance of limited activities on a treatment table. Intensive emphasis on a specific pattern of facilitation during treatment on a table may produce a better gait pattern than will practice of a less adequate pattern during walking. Resisted planti-grade walking on a mat may more readily achieve lengthening of posterior structures than will re-peated and laborious attempts to relax hamstring muscles when the patient is lying supine on a table. Thus, integration of activities requires that goals be established, that the proper time and the proper place be determined for emphasis of those activities which will hasten the patient's motor learning and will integrate his neuromuscular abilities.

Integration of activities must extend beyond the physical therapy program (ref. 29). In-bed activities on the ward, occupational therapy projects, recreational activities, and any other activity must be directed toward the goals established for the individual patient. Clearly defined long-term goals give direction. Specifically described short-term goals have the same direction but are phases of advancement. Long-term goals may have to be altered, raised, or lowered, as the patient's potentialities become clearer. Short-term goals require constant alteration in keeping with the patient's status and his progress. A goal is a challenge to the patient and to the staff.

6. References and suggested reading

References

1. Buchwald, J. S. "Exteroceptive Reflexes and Movement," in: *Proceedings of an Exploratory and Analytical Survey of Therapeutic Exercise (NU-STEP),* Northwestern University Medical School, July 25–August 19, 1966. *Amer. J. Phys. Med. 46:*121–128, 1967.
2. Dorland, W. A. N. *The American Illustrated Medical Dictionary,* 24th ed., Philadelphia, Saunders, 1965.
3. Freeman, J. T. Posture in the aging and aged body. *J.A.M.A. 165:*843–846, 1957.
4. Geldard, F. A. Some neglected possibilities of communication. *Science 131:*1583–1588, 1960.
5. Gellhorn, E. Patterns of muscular activity in man. *Arch. Phys. Med. 28:9:*568–574, 1947.
6. Gesell, A., and Amatruda, C. S. *Developmental Diagnosis,* 2nd ed. New York, Hoeber, 1947.
7. Gray, H. *Anatomy of the Human Body,* ed. by Goss, C. M. 27th ed. Philadelphia, Lea & Febiger, 1959, pp. 32–46.
8. Hagbarth, K. E. Excitatory and inhibitory skin areas for flexor and extensor motoneurones. *Acta. Physiol. Scand. 26* (Suppl. 94): 1–58, 1952.
9. Harrison, V. F. A review of the neuromuscular bases for motor learning. *Research Quarterly 33:*59–69, 1962.
10. Hellebrandt, F. A. "Physiology," in DeLorme, T. L., and Watkins, A. L., *Progressive Resistance Exercise,* Chapter 2. New York, Appleton-Century-Crofts, 1951.
11. Hellebrandt, F. A. Application of the overload principle to muscle training in man. *Amer. J. Phys. Med. 37:*278–283, 1958.
12. Hellebrandt, F. A., Houtz, S. J., and Eubank, R. N. Influence of alternate and reciprocal exercise on work capacity. *Arch. Phys. Med. 32:*766–776, 1951.
13. Hellebrandt, F. A., Schade, M., and Carns, M. L. Methods of evoking the tonic neck reflexes in normal human subjects. *Amer. J. Phys. Med. 41:*90–139, 1962.
14. Hellebrandt, F. A., and Waterland, J. C. Indirect learning: The influence of unimanual exercise on related muscle groups of the same and opposite side. *Amer. J. Phys. Med. 41:*45–55, 1962.
15. Hooker, D. *The Prenatal Origin of Behavior.* Porter Lectures, Series 18. Lawrence, Kansas, Universty of Kansas Press, 1952.
16. Humphrey, T. The trigeminal nerve in relation to early human fetal activity. *Res. Publ. Ass. Nerv. Ment. Dis. 33:*127–154, 1954.
17. Jacobs, M. "Developmental Basis for Therapeutic Exercise," in *Proceedings of the Third International Congress, 1959, World Confederation for Physical Therapy,* Paris, French Committee, W.C.P.T., 1961.
18. Kabat, H. "Proprioceptive Facilitation in Therapeutic Exercise," in *Therapeutic Exercise,* ed. by Licht, S. 2nd ed., Chapter 13. New Haven, E. Licht, 1961.
19. Kabat, H., and Knott, M. Proprioceptive facilitation technics for treatment of paralysis. *Phys. Ther. Rev. 33:*53–64, 1953 .
20. Levine, M. G., Kabat, H., Knott, M., and Voss, D. E. Relaxation of spasticity by physiological technics. *Arch. Phys. Med. 35:*214–223, 1954.
21. Levine, M. G., Knott, M., and Kabat, H. Relaxation of spasticity by electrical stimulation of antagonistic muscles. *Arch. Phys. Med. 33:*668–673, 1952.
22. McGraw, M. B. *The Neuromuscular Maturation of the Human Infant.* New York, Columbia University Press, 1943. Reprinted edition, New York, Hafner Publishing Company, 1962.
23. Mead, S. Personal communication, 1963.
24. Peele, T. L. *The Neuroanatomical Basis for Clinical Neurology.* New York, McGraw-Hill, 1954.
25. Robinson, M. E., Doudlah, A. M., and Waterland, J. C. The influence of vision on the performance of a motor act. *Amer. J. Occup. Ther. 19:*202–204, 1965.
26. Rood, M. S. Neurophysiological mechanisms utilized in the treatment of neuromuscular dysfunction. *Amer. J. Occup. Ther. 10:*220–224, 1956.
27. Sherrington, C. *The Integrative Action of the Nervous System.* New Haven, Yale University Press, reprinted ed. 1961, p. 340.
28. Toussaint, D., and Knott, M. The use of wall pulleys with mat activities. *Phys. Ther. Rev. 35:*477–483, 1955.
29. Voss, D. E. Proprioceptive neuromuscular facilitation: Application of patterns and techniques in occupational therapy. *Amer. J. Occup. Ther. 13:*191–194, 1959.

Suggested reading

Note: Documentation of the text as revealed by the list of references, is limited. Complete documentation would have yielded references perhaps more lengthy than the text itself. Thus, many writings that could have been used for documentation appear in the Suggested Readings; for example, Buchwald's *Basic Mechanisms of Motor Learning,* and Eldred's *Posture and Locomotion.*

Overlapping is a characteristic in scientific writings, thus Suggested Readings in neurophysiology, motor development, and motor learning support the concept that behavior is a function of structure (Ames). Anatomy is a science of structure; physiology is a science of function. An understanding of both is necessary to an understanding of behavior.

The Proceedings of an Exploratory and Analytical Survey of Therapeutic Exercise, held at Northwestern University Medical School, July 25–August 19, 1966, has been published in the *American Journal of Physical Medicine,* Vol. 46, no. 1, 1967. The Proceedings includes a galaxy of papers by Buchwald, Eldred, Fischer, Jacobs, Walters, and Waterland, and others whose work does not appear in our lists. Readers are urged to consult and study the Proceedings for recent material on neurophysiology, motor development, and motor learning.

BASIC INFORMATION

Neurophysiology

* Bouman, H. D. Some considerations of muscle activity. *J. Amer. Phys. Ther. Ass. 45:*431–436, 1965.
* Bouman, H. D. Some considerations of the physiology of sensation. *J. Amer. Phys. Ther. Ass. 45:*573–577, 1965.

Denny-Brown, D. "Motor mechanisms—Introduction: The General Principles of Motor Integration," in *Handbook of Physiology, Section 1: Neurophysiology,* Vol. II, edited by Field, J., Magoun, H. W., and Hall, V. E., Chapter 32. Washington, D.C., American Physiological Society, 1960.

* Eldred, E. The dual sensory role of muscle spindles. *J. Amer. Phys. Ther. Ass. 45:*290–313, 1965.

* Reprinted in: *The Child with Central Nervous System Deficit,* Washington, D.C., Department of Health Education and Welfare, Children's Bureau Publication No. 432–1965.

* Eldred, E. Postural integration at spinal levels. *J. Amer. Phys. Ther. Ass. 45:*332–344, 1965.

Eldred, E. "Posture and Locomotion," in *Handbook of Physiology, Section 1: Neurophysiology,* Vol. II, edited by Field, J., Magoun, H. W., and Hall, V. E., Chapter 41, Washington, D.C., American Physiological Society, 1960.

Fischer, E. Neurophysiology a physical therapist should know. *Phys. Ther. Rev. 38:*741–748, 1958.

Fischer, E. Physiological basis of methods to elicit, reinforce, and coordinate muscle movement. *Phys. Ther. Rev. 38:*468–473, 1958.

Fischer, E. Physiological basis of volitional movement. *Phys. Ther. Rev. 38:*405–412, 1958.

Granit, R. *Receptors and Sensory Perception.* New Haven, Yale University Press, 1962. (Yale paperbound edition.)

Harrison, V. F. Review of skeletal muscle. Review of sensory receptors in skeletal muscles with special emphasis on the muscle spindle. Review of motor unit. *Phys. Ther. Rev. 41:*17–40, 1961.

Paillard, J. "The Patterning of Skilled Movement," in *Handbook of Physiology, Section 1: Neurophysiology,* Vol. III edited by Field, J., Magoun, H. W., and Hall, V. E., Chapter 67. Washington, D.C., American physiological Society, 1960.

Ralston, H. J. Recent advances in nueromuscular physiology. *Amer. J. Phys. Med. 36:*94–120, 1957.

Ralston, H. J. Some considerations of the physiological basis of therapeutic exercise. *Phys. Ther. Rev 38:*465–468, 1958.

* Twitchell, T. E. Attitudinal reflexes. *J. Amer. Phys. Ther. Ass. 45:*411–418, 1965.

Motor Development

Ames, L. B. Individuality of motor development. *J. Amer. Phys. Ther. Ass. 46:*121–127, 1966.

Gesell, A. "Behavior Patterns of Fetal-Infant and Child," in *Genetics and the Inheritance of Integrated Neurological and Psychiatric Patterns. Proc. Ass. Research Nerv. & Ment. Dis. 33:*114–123, 1954.

Gesell, A., and Amatruda, C. S. *The Embryology of Behavior.* New York, Harper, 1945.

Hellebrandt, F. A., Rarick, L., Glasgow, R., and Carns, M. L. Physiological analysis of basic motor skills. I. Growth and development of jumping. *Amer. J. Phys. Med. 40:*14–25, 1961.

Monie, I. W. Development of motor behavior. *J. Amer. Phys. Ther. Ass. 43:*333–338, 1963.

* Twitchell, T. E. Normal motor development. *J. Amer. Phys. Ther. Ass. 45:*419–423, 1965.

* Twitchell, T. E. Variations and abnormalities of motor development. *J. Amer. Phys. Ther. Ass. 45:*424–430, 1965.

Weisz, S. Studies in equilibrium reaction. *J. Nerv. Ment. Dis. 88:*150–162, 1938.

Motor Learning

* Buchwald, J. S. Basic mechanisms of motor learning. *J. Amer. Phys. Ther. Ass.* 45:314–331, 1965.

Forward, E. Implications of research in motor learning for physical therapy. *J. Amer. Phys. Ther. Ass.* 43:339–344, 1963.

Hellebrandt, F. A. Physiology of motor learning as applied to the treatment of the cerebral palsied. *Quart. Rev. Pediat.* 7:5–14, 1952.

Hellebrandt, F. A. Kinesthetic awareness of motor learning. *Cerebral Palsy Rev.* 14:5–6, 1953.

Hellebrandt, F. A. The physiology of motor learning. *Cerebral Palsy Rev.* 19:9–14, 1958.

Hellebrandt, F. A., Parrish, A. M., and Houtz, S. J. Cross education: The influence of unilateral exercise on the contralateral limb. *Arch. Phys. Med.* 28:76–85, 1947.

Walters, C. E. The effect of overload on bilateral transfer of a motor skill. *Phys. Ther. Rev.* 35:567–569, 1955.

Further Studies on Normal Subjects

Hellebrandt, F. A. Cross education: Ipsilateral and contralateral effects of unimanual training. *J. Appl. Physiol.* 4:136–144, 1951.

Hellebrandt, F. A., and Houtz, S. J. Mechanisms of muscle training: The influence of pacing. *Phys. Ther. Rev.* 38:319–322, 1958.

Hellebrandt, F. A., Hockman, D. E., and Partridge, M. J. Physiological effects of simultaneous static and dynamic exercise. *Amer. J. Phys. Med.* 35:106–117, 1956.

Hellebrandt, F. A., Houtz, S. J., and Krikorian, A. M. Influence of bimanual exercise on unilateral work capacity. *J. Appl. Physiol.* 2:446–452, 1950.

Hellebrandt, F. A., Houtz, S. J., Partridge, M. J., and Walters, C. E. Tonic neck reflexes in exercises of stress in man. *Amer. J. Phys. Med.* 35:144–159, 1956.

Hellebrandt, F. A., and Waterland, J. C. Expansion of motor patterning under exercise stress. *Amer. J. Phys. Med.* 41:56–66, 1962.

Latimer, R. Utilization of tonic and labyrinthine reflexes for the facilitation of work output. *Phys. Ther. Rev.* 33:237–241, 1953.

Murray, M. P., Drought, A. B., and Kory, R. C. Walking patterns of normal men. *J. Bone Joint Surg.* 46-A:335–360, 1964.

O'Connell, A. L. Electromyographic study of certain leg muscles during movements of the free foot and during standing. *Amer. J. Phys. Med.* 37:289–301, 1958.

Partridge, M. J. Electromyographic demonstration of facilitation. *Phys. Ther. Rev.* 34:227–233, 1954.

Waterland, J. C., and Hellebrandt, F. A. Involuntary patterning associated with willed movement performed against progressively increasing resistance. *Amer. J. Phys. Med.* 43:13–30, 1964.

Waterland, J. C., and Munson, N. Involuntary patterning evoked by exercise stress. *J. Amer. Phys. Ther. Ass.* 44:91–97, 1964.

Waterland, J. C., and Munson, N. Reflex association of head and shoulder girdle in nonstressful movements in man. *Amer. J. Phys. Med.* 43:98–108, 1964.

Wellock, L. M. Development of bilateral muscular strength through ipsilateral exercise. *Phys. Ther. Rev.* 38:671–675, 1958.

RELATED INFORMATION

General

Kabat, H. Central mechanisms for recovery of neuromuscular function. *Science* 112:2897:23–24, 1950.

Kabat, H. The role of central facilitation in restoration of motor function in paralysis. *Arch. Phys. Med.* 33:521–533, 1952.

Levine, M. G., and Kabat, H. Proprioceptive facilitation of voluntary motion in man. *J. Nerv. Ment. Dis.* 117:199–211, 1953.

Voss, D. E., and Knott, M. Patterns of motion for proprioceptive neuromuscular facilitation. *Brit. J. Phys. Med.* 17:191–198, 1954.

Clinical Applications

Ault, M. M. Facilitation tecnics used to relieve contractures in a rheumatoid arthritis patient. *Phys. Ther. Rev.* 40:657–658, 1960.

Chrystal, M., and Rosner, H. Mass movement patterns in neuromuscular reeducation. *Phys. Ther. Rev.* 34:344–345, 1954.

Ionta, M. K. Facilitation technics in the treatment of early rheumatoid arthritis. *Phys. Ther. Rev.* 40:119–120, 1960.

Kabat, H. Restoration of function through neuromuscular reeducation in traumatic paraplegia. *A.M.A. Arch. Neurol. Psychiat.* 67:737–744, 1952.

Kabat, H. Analysis and therapy of cerebellar ataxia and asynergia. *A.M.A. Arch. Neurol. Psychiat.* 74:375–382, 1955.

Kabat, H., and McLeod, M. Athetosis: Neuromuscular dysfunction and treatment. *Arch. Phys. Med.* 40:285–292, 1959.

Kabat, H., and McLeod, M. Neuromuscular dysfunction and treatment of athetosis. *Physiotherapy* 46:125–129, 1960.

Kabat, H., McLeod, M., and Holt, C. Neuromuscular dysfunction and treatment of corticospinal lesions. *Physiotherapy* 45:251–257, 1959.

Knott, M. Specialized neuromuscular technics in the treatment of cerebral palsy. *Phys. Ther. Rev.* 32:73–75, 1952.

Knott, M. Report of a case of Parkinsonism treated with proprioceptive facilitation technics. *Phys. Ther. Rev.* 37:229, 1957.

Knott, M. Avulsion of a finger with protracted disability. *Phys. Ther. Rev.* 38:552, 1958.

Knott, M. Bulbar involvement with good recovery. *J. Amer. Phys. Ther. Ass.* 42:38–39, 1962.

Knott, M. Neuromuscular facilitation in the treatment of rheumatoid arthritis. *J. Amer. Phys. Ther. Ass.* 44:737–739, 1964.

Knott, M. Neuromuscular facilitation in the child with central nervous system deficit. *J. Amer. Phys. Ther. Ass.* 46:721–724, 1966.

Knott, M., and Barufaldi, D. Treatment of whiplash injuries. *Phys. Ther. Rev.* 41:573–577, 1961.

Knott, M., and Mead, S. Facilitation technics in lower extremity amputations. *Phys. Ther. Rev.* 40:587–589, 1960.

Nunley, R. L., and Bedini, S. J. Paralysis of the shoulder subsequent to a comminuted fracture of the scapula. *Phys. Ther. Rev.* 40:442–447, 1960.

Torp, M. J. Adaptations of neuromuscular facilitation technics. *Phys. Ther. Rev.* 36:577–586, 1956.

Torp, M. J. An exercise program for the brain-injured. *Phys. Ther. Rev.* 36:644–675, 1956.

Toussaint, D. Facilitation technics achieve self-care in poliomyelitis patient. *Phys. Ther. Rev.* 37:590, 1957.

Voss, D. E. "Proprioceptive Neuromuscular Facilitation: Demonstrations with Cerebral Palsied Child, Hemiplegic Adult, Arthritic Adult, Parkinsonian Adult," in *Exploratory and Analytical Survey of Therapeutic Exercise (NU-STEP)*, Northwestern University Medical School, July 25–August 19, 1966. *Amer. J. Phys. Med.* 46:838–898, 1967.

Voss, D. E., Knott, M., and Kabat, H. Application of neuromuscular facilitation in the treatment of shoulder disabilities. *Phys. Ther. Rev.* 33:536–541, 1953.

Adjuncts to Facilitation Techniques

Bassett, S. W., and Lake, B. M. Use of cold applications in the management of spasticity. *Phys. Ther. Rev. 38:333–334*, 1958.

Boes, M. C. Reduction of spasticity by cold. *J. Amer. Phys. Ther. Ass. 42:29–32*, 1962.

Boynton, B. L., Garramone, P. M., and Buca, J. T. Observations on the effects of cool baths for patients with multiple sclerosis. *Phys. Ther. Rev. 39:297–299*, 1959.

Conway, B. Ice packs in diabetic neuropathy. *Phys. Ther. Rev. 41:586–588*, 1961.

Davies, E. J., Perry, J. H., and Wakefield, P. "Afferent Stimuli to Facilitate or Inhibit Motor Activity." (Techniques Developed by M. Rood.) In *Motor Integration*, ed. by Decker, R., Springfield, Illinois, Charles C Thomas, 1962, Chapter 5, pp. 73–83.

Liberson, W. T. Experiment concerning reciprocal inhibition of antagonists elicited by electrical stimulation of agonists in a normal individual. *Amer. J. Phys. Med. 44: 306–308*, 1965.

Lorenze, E. J., Carantonis, G., and De Rosa, A. J. Effect on coronary circulation of cold packs to hemiplegic shoulders. *Arch Phys. Med. 41:394–399*, 1960.

Miglietta, O. E. Evaluation of cold in spasticity. *Amer. J. Phys. Med. 41:148–151*, 1962.

Petajan, J. H., and Watts, N. Effects of cooling on the triceps surae muscle. *Amer. J. Phys. Med. 41:240–251*, 1962.

Rockefeller, L. E. The use of cold packs for increasing joint range of motion. *Phys. Ther. Rev. 38:564–566*, 1958.

Viel, E. Treatment of spasticity by exposure to cold. *Phys. Ther. Rev. 39:598–599*, 1959.

Walters, C. E., Garrison, L., Duncan, H. J., Hopkins, F. V., and Snyder, J. W. The effects of therapeutic agents on muscular strength and endurance. *Phys. Ther. Rev. 40: 266–270*, 1960.

Watson, C. W. Effect of lowering body temperature on the symptoms and signs of multiple sclerosis. *New Engl. J. Med. 261:1253–1259*, 1959.

Supportive Comments

Gillette, H. E. Changing concepts in the management of neuromuscular dysfunction. *Southern Med. J. 52:1227–1229*, 1959.

Mead, S. "A Six-Year Evaluation of Proprioceptive Neuromuscular Facilitation Technics," in *Proceedings of the Third International Congress of Physical Medicine, 1960*. Chicago, American Congress of Physical Medicine and Rehabilitation and American Academy of Physical Medicine and Rehabilitation, 1962.

Watkins, A. L. Medical progress: Physical medicine and rehabilitation. *New Engl. J. Med. 255:1233–1239*, 1956.

7. Reference tables

Pattern combinations for reinforcement (Tables 3-9)

Table 3. Head and Neck Patterns Reinforced by Pattern Combinations of Upper Extremities

Pattern to be reinforced	Upper extremity patterns (eye follows hand)
1. Head and neck flexion with rotation (left/right)	Extension–adduction–internal rotation (contralateral) Bilateral assymetrical (chopping) (ipsilateral)
2. Head and neck extension with rotation (left/right)	Flexion–abduction–external rotation (ipsilateral) Bilateral assymetrical (lifting) (ipsilateral)
3. Head and neck rotation (left/right)	Extension–abduction–internal rotation (ipsilateral) Flexion–adduction–external rotation (contralateral)

Table 4. Upper Trunk Patterns Reinforced by Pattern Combinations of Head and Neck, Lower Trunk, and Upper Extremities

Pattern to be reinforced	Neck patterns (ipsilateral)	Lower trunk patterns	Upper extremity patterns
1. Upper trunk flexion with rotation (left/right)	Head and neck flexion with rotation (ipsilateral)	1. Lower trunk flexion with rotation (ipsilateral) Flexion–abduction–internal rotation (ipsilateral lower extremity) Flexion–adduction–external rotation (contralateral lower extremity) 2. Lower trunk flexion with rotation (contralateral) Flexion–abduction–internal rotation (contralateral lower extremity) Flexion–adduction–external rotation (ipsilateral lower extremity)	1. Bilateral asymmetrical patterns (chopping, hands approximated) ipsilateral Extension–adduction–internal rotation (contralateral upper extremity) Extension–abduction–internal rotation (ipsilateral upper extremity) 2. Unilateral upper extremity patterns Extension–adduction–internal rotation (contralateral upper extremity)
1. Upper trunk extension with rotation (left/right)	Head and neck extension with rotation (ipsilateral)	1. Lower trunk extension with rotation (ipsilateral) Extension–abduction–internal rotation (ipsilateral lower extremity) Extension–adduction–external rotation (contralateral lower extremity) 2. Lower trunk extension with rotation (contralateral) Extension–abduction–internal rotation (contralateral lower extremity) Extension–adduction–external rotation (ipsilateral lower extremity) *Note:* All lower trunk patterns may be performed with bilateral knee flexion, knee extension, or the knees may remain straight. Extremities are held in close approximation.	1. Bilateral asymmetrical patterns (lifting, hands approximated) ipsilateral Flexion–abduction–external rotation (ipsilateral upper extremity) Flexion–adduction–external rotation (contralateral upper extremity) 2. Unilateral upper extremity pattern Flexion–abduction–external rotation (ipsilateral upper extremity) *Note:* Since neck patterns are the key to the upper trunk patterns, when upper extremity patterns are used to reinforce the upper trunk, the eyes follow the hands.
3. Upper trunk rotation (left/right)	Head and neck rotation (ipsilateral)	1. Lower trunk extension with rotation (ipsilateral) 2. Lower trunk flexion with rotation (contralateral)	1. Bilateral reciprocal patterns Extension–abduction–internal rotation (ipsilateral) Flexion–adduction–external rotation (contralateral) 2. Unilateral upper extremity patterns Extension–abduction–internal rotation (ipsilateral) Flexion–adduction–external rotation (contralateral)

Table 5. Pattern Combinations for Reinforcement of Lower Trunk Patterns

Pattern to be reinforced	Head and neck and upper trunk patterns	Bilateral asymmetrical upper extremity patterns	Unilateral upper extremity patterns
1. Lower trunk flexion with rotation (left/right)	Flexion with rotation (ipsilateral) Flexion with rotation (contralateral)	1. Chopping (ipsilateral) 2. Chopping (contralateral)	1. Flexion–adduction–external rotation (contralateral) 2. Extension–adduction–internal rotation (ipsilateral)
2. Lower trunk extension with rotation (left/right)	Extension with rotation (ipsilateral) Extension with rotation (contralateral)	1. Lifting (ipsilateral) 2. Lifting (contralateral)	1. Extension–abduction–internal rotation (ipsilateral) 2. Flexion–abduction–external rotation (contralateral)

Table 6. Upper Extremity Patterns Reinforced by Pattern Combinations of Head and Neck and Lower Extremities

Pattern to be reinforced	Head and neck patterns (eye follows hand)	Ipsilateral or contralateral lower extremity patterns
1. Flexion–adduction–external rotation	Extension with rotation (contralateral) Rotation (contralateral)	Flexion–adduction–external rotation Flexion–abduction–internal rotation
2. Extension–abduction–internal rotation	Flexion with rotation (ipsilateral) Rotation (ipsilateral)	Extension–abduction–internal rotation Extension–adduction–external rotation
3. Flexion–abduction–external rotation	Extension with rotation (ipsilateral)	Extension–adduction–external rotation Extension–abduction–internal rotation
4. Extension–adduction–internal rotation	Flexion with rotation (contralateral) Rotation (contralateral)	Flexion–adduction–external rotation Flexion–abduction–internal rotation

Table 7. Lower Extremity Patterns Reinforced by Pattern Combinations of Head and Neck and Upper Extremities

Pattern to be reinforced	Head and neck patterns	Ipsilateral or contralateral upper extremity
1. Flexion–adduction–external rotation	Flexion with rotation (ipsilateral) Rotation (ipsilateral)	Flexion–adduction–external rotation Extension–adduction–internal rotation
2. Extension–abduction–internal rotation	Extension with rotation (ipsilateral) Rotation (ipsilateral)	Extension–abduction–internal rotation Flexion–abduction–external rotation
3. Flexion–abduction–internal rotation	Flexion with rotation (ipsilateral) Rotation (ipsilateral)	Flexion–adduction–external rotation Extension–adduction–internal rotation
4. Extension–adduction–external rotation	Extension with rotation (contralateral) Rotation (contralateral)	Extension–abduction–internal rotation Flexion–abduction–internal rotation

Table 8. Upper Extremity Pattern Combinations for Reinforcement of Opposite Upper Extremity

Pattern to be reinforced	Bilateral symmetrical	Bilateral asymmetrical	Bilateral reciprocal (same diagonal)	Bilateral reciprocal (opposite diagonal)
1. Flexion–adduction–external rotation	Flexion–adduction–external rotation	Flexion–abduction–external rotation	Extension–abduction–internal rotation	Extension–adduction–internal rotation
2. Extension–abduction–internal rotation	Extension–abduction–internal rotation	Extension–adduction–internal rotation	Flexion–adduction–external rotation	Flexion–abduction–external rotation
3. Flexion–abduction–external rotation	Flexion–abduction–external rotation	Flexion–adduction–external rotation	Extension–adduction–internal rotation	Extension–abduction–internal rotation
4. Extension–adduction–internal rotation	Extension–adduction–internal rotation	Extension–abduction–internal rotation	Flexion–abduction–external rotation	Flexion–adduction–external rotation

Table 9. Lower Extremity Pattern Combinations for Reinforcement of Opposite Lower Extremity

Pattern to be reinforced	Bilateral symmetrical	Bilateral asymmetrical	Bilateral reciprocal (same diagonal)	Bilateral reciprocal (opposite diagonal)
1. Flexion–adduction–external rotation	Flexion–adduction–external rotation	Flexion–abduction–internal rotation	Extension–abduction–internal rotation	Extension–adduction–external rotation
2. Extension–abduction–internal rotation	Extension–abduction–internal rotation	Extension–adduction–external rotation	Flexion–adduction–external rotation	Flexion–abduction internal rotation
3. Flexion–abduction–internal rotation	Flexion–abduction–internal rotation	Flexion–adduction–external rotation	Extension–adduction–external rotation	Extension–abduction–internal rotation
4. Extension–adduction–external rotation	Extension–adduction–external rotation	Extension–abduction–internal rotation	Flexion–abduction–internal rotation	Flexion–adduction–external rotation

Optimal patterns for individual muscles (Tables 10-13)

Table 10. Optimal Patterns for Muscles of the Head and Neck

Muscles (left muscles considered)	Patterns
Platysma	Flexion with rotation—left
Trapezius	Extension with rotation—left
Levator scapulae	Rotation—left
Sternocleidomastoideus	Flexion with rotation—left—lengthened to middle range Flexion with rotation—right—middle to shortened range Rotation—left—lengthened to middle range Rotation—right—middle to shortened range
Suprahyoidei Infrahyoidei	Flexion with rotation—left
Rectus capitis lateralis	Flexion with rotation—right
Rectus capitis anterior	Flexion with rotation—left
Longus colli Longus capitis	Flexion with rotation—left
Scalenus anterior Scalenus medius Scalenus posterior	Flexion with rotation—left Rotation—left
Rectus capitis posterior minor Rectus capitis posterior major Obliquus capitis inferior Obliquus capitis superior	Extension with rotation—left
Splenius capitis Longissimus capitis Splenius cervicis Longissimus cervicis Iliocostalis cervicis Interspinales Intertransversarii Semispinalis capitis	Extension with rotation—left
Semispinalis cervicis Multifidus	Extension with rotation—right

Table 11. Optimal Patterns for Muscles of the Trunk

Muscles (left muscles considered)	Patterns
Spinalis thoracis Longissimus thoracis Iliocostalis thoracis Iliocostalis lumborum Sacrospinalis Interspinales Intertransversarii	Trunk extension with rotation—left
Semispinalis thoracis Multifidus Rotatores	Trunk extension with rotation—right
Quadratus lumborum	Trunk extension with rotation—left Trunk flexion with rotation—left Trunk rotation—left
Obliquus externus	Upper trunk flexion with rotation—right Lower trunk flexion with rotation—left
Obliquus internus	Upper trunk flexion with rotation—left Lower trunk flexion with rotation—right
Rectus abdominis—left portion	Upper trunk flexion with rotation—left Lower trunk flexion with rotation—left
Transversus abdominis	Trunk extension with rotation—left Upper trunk rotation—left
Intercostales externi Serratus posterior superior Diaphragma—descent of dome	Upper trunk extension with rotation—left
Levators costarum Serratus posterior inferior	Upper trunk extension with rotation—right
Intercostales interni Subcostales Diaphragma—ascent of dome	Upper trunk flexion with rotation—left
Tranversus thoracis	Upper trunk flexion with rotation—right

Table 12. Optimal Patterns for Muscles of Upper Extremity with Consideration for Action on Two or More Joints

Muscles	Patterns
Shoulder girdle	
Serratus anterior	Flexion–adduction–external rotation
Levator scapulae Rhomboideus major Rhomboideus minor Latissimus dorsi—inferior angle attachment	Extension–abduction–internal rotation
Trapezius	Flexion–abduction–external rotation
Subclavius Pectoralis minor	Extension–adduction–internal rotation
Pectoralis major—clavicular portion Deltoideus—anterior portion Coracobrachialis	Flexion–adduction–external rotation
Deltoideus—posterior portion Teres Major Latissimus dorsi	Extension–abduction–internal rotation
Latissimus dorsi–shortened range	Extension–adduction–internal rotation-posteriorly
Supraspinatus Infraspinatus Teres minor Deltoideus—middle portion	Flexion–abduction–external rotation
Pectoralis major—sternal portion Subscapularis	Extension–adduction–internal rotation
Elbow	
Biceps brachii Brachialis	Flexion–adduction–external rotation with elbow flexion
Triceps brachii Anconeus Subanconeus	Extension–abduction–internal rotation with elbow extension
Forearm	
Supinator	Flexion–adduction–external rotation
Pronator quadratus Brachioradialis	Extension–abduction–internal rotation Flexion–abduction–external rotation with elbow flexion
Pronator teres	Extension–adduction–internal rotation with elbow flexion
Wrist	
Flexor carpi radialis	Flexion–adduction–external rotation with elbow flexion
Extensor carpi ulnaris	Extension–abduction–internal rotation with elbow extension
Palmaris longus	Flexion–adduction–external rotation with elbow flexion Extension–adduction–internal rotation with elbow flexion
Flexor carpi ulnaris	Extension–adduction–internal rotation with elbow flexion
Extensor carpi radialis longus Extensor carpi radialis brevis	Flexion–abduction–external rotation with elbow extension
Hand and fingers	
Flexor digitorum superficialis	Flexion–adduction–external rotation with elbow flexion Extension–adduction–internal rotation with elbow flexion
Flexor digitorum profundus	Flexion–adduction–external rotation Extension–adduction–internal rotation

Table 12. Optimal Patterns for Muscles of Upper Extremity with Consideration for Action on Two or More Joints (Continued)

Muscles	Patterns
Interossei palmares	Flexion–adduction–external rotation Extension–adduction–internal rotation
Flexor digiti minimi brevis	Flexion–adduction–external rotation
Opponens digiti minimi	Flexion–adduction–external rotation
Extensor digitorum communis	Flexion–abduction–external rotation with elbow extension Extension–abduction–internal rotation with elbow extension
Interossei dorsales	Flexion–abduction–external rotation Extension–abduction–internal rotation
Extensor indicis proprius	Flexion–abduction–external rotation
Extensor digiti minimi	Extension–abduction–internal rotation
Abductor digiti minimi	Extension–abduction–internal rotation
Lumbricales	All patterns of upper extremity

Thumb

Muscles	Patterns
Flexor pollicis longus Flexor pollicis brevis Adductor pollicis	Flexion–adduction–external rotation
Abductor pollicus brevis	Extension–abduction–internal rotation
Abductor pollicus longus Extensor pollicis longus Extensor pollicis brevis First interosseus dorsalis	Flexion–abduction–external rotation
Opponens pollicis Palmaris brevis	Extension–adduction–internal rotation

Table 13. Optimal Patterns for Muscles of Lower Extremity with Consideration for Action on Two or More Joints

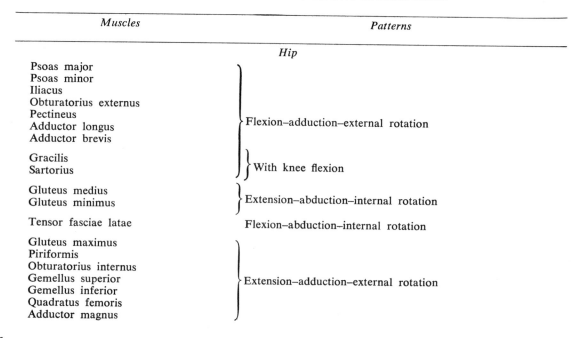

Muscles	Patterns
Hip	
Psoas major Psoas minor Iliacus Obturatorius externus Pectineus Adductor longus Adductor brevis	Flexion–adduction–external rotation
Gracilis Sartorius	With knee flexion
Gluteus medius Gluteus minimus	Extension–abduction–internal rotation
Tensor fasciae latae	Flexion–abduction–internal rotation
Gluteus maximus Piriformis Obturatorius internus Gemellus superior Gemellus inferior Quadratus femoris Adductor magnus	Extension–adduction–external rotation

Table 13. Optimal Patterns for Muscles of Lower Extremity with Consideration for Action on Two or More Joints (Continued)

Muscles	Patterns
Knee	
Rectus femoris—medial portion	Flexion–adduction–external rotation with knee extension
Vastus medialis	Extension–adduction–external rotation with knee extension Flexion–adduction–external rotation with knee extension
Biceps femoris Popliteus	Extension–abduction–internal rotation with knee flexion Flexion–abduction–internal rotation with knee flexion
Rectus femoris—lateral portion	Flexion–abduction–internal rotation with knee extension
Vastus intermedius Vastus lateralis	Extension–abduction–internal rotation with knee extension Flexion–abduction–internal rotation with knee extension
Semitendinosus Semimembranosus	Extension–adduction–external rotation with knee flexion Flexion–adduction–external rotation with knee flexion
Articularis genus	All patterns with knee extension
Ankle and foot	
Tibialis anterior	Flexion–adduction–external rotation
Peroneus longus Gastrocnemius—lateral portion Soleus—lateral portion	Extension–abduction–internal rotation
Peroneus brevis Peroneus tertius	Flexion–abduction–internal rotation
Tibialis posterior Gastrocnemius—medial portion Soleus—medial portion Plantaris	Extension–adduction–external rotation
Foot and toes	
Extensor hallucis longus Extensor digitorum longus Extensor digitorum brevis Interossei dorsales	Flexion–abduction–internal rotation Flexion–adduction–external rotation
Flexor hallucis longus Flexor digitorum longus Flexor digitorum brevis Flexor hallucis brevis Interossei plantares	Extension–adduction–external rotation Extension–abduction–internal rotation
Flexor digiti minimi brevis Adductor hallucis Quadratus plantae—lateral portion	Extension–abduction–internal rotation
Quadratus plantae—medial portion	Extension–adduction–external rotation
Lumbricales	All patterns of lower extremity

Optimal patterns according to peripheral innervation (Tables 14-15)

Table 14. Optimal Patterns for Muscles of Upper Extremity According to Peripheral Innervation

Nerve	Flexion–adduction–external rotation	Extension–abduction–internal rotation	Flexion–abduction–external rotation	Extension–adduction–internal rotation
Spinal accessory and C3–4				
Trapezius	– – –	– – –	x x x	– – –
Dorsal scapular C3–4,				
Levator scapulae	– – –	x x x	– – –	– – –
Dorsal scapular C5				
Rhomboideii major and minor	– – –	x x x	– – –	– – –
Suprascapular nerve C5–6				
Supraspinatus	– – –	– – –	x x x	– – –
Infraspinatus	– – –	– – –	x x x	– – –
Subclavius C5–6				
Subclavius	– – –	– – –	– – –	x x x
Subscapular nerves C5–6				
Subscapularis	– – –	– – –	– – –	x x x
Teres major	– – –	x x x	– – –	– – –
Long thoracic nerve C5–6–7				
Serratus anterior	x x x	– – –	– – –	– – –
Axillary nerve, C5–6				
Deltoid	x x x	x x x	x x x	– – –
Teres minor	– – –	– – –	x x x	– – –
Musculocutaneous nerve, C5–6–7	x x x			
Coracobrachialis	x x x	– – –	– – –	– – –
Biceps	x x x Elbow flexion	– – –	– – –	– – –
Brachialis	x x x Elbow flexion	– – –	– – –	– – –
Anterior thoracic-lateral C5–6–7				
Pectoralis major —clavicular	x x x	– – –	– – –	– – –
Anterior thoracic-medial C8–T1				
Pectoralis minor	– – –	– – –	– – –	x x x
Pectoralis major —sternal	– – –	– – –	– – –	x x x
Thoracodorsal nerve C6–7–8				
Latissimus dorsi	– – –	x x x	– – –	Posteriorly with elbow flexion
Radial nerve—C6–7–8				
Triceps brachii	– – –	x x x Elbow extension	– – –	– – –

Note: Scapular motions are more easily controlled if the elbow remains straight. Elbow flexion may be allowed in the flexion patterns; elbow extension may be allowed in the extension patterns.

Table 14. Optimal Patterns for Muscles of Upper Extremity According to Peripheral Innervation (Continued)

Nerve	Flexion–adduction–external rotation	Extension–abduction–internal rotation	Flexion–abduction–external rotation	Extension-adduction–internal rotation
Brachioradialis	– – –	– – –	x x x Elbow flexion	– – –
Extensor carpi radialis longus	– – –	– – –	x x x	– – –
Anconeus	– – –	x x x Elbow extension	– – –	– – –
Extensor carpi radialis brevis	– – –	– – –	x x x	– – –
Extensor digitorum communis	– – –	x x x	x x x	– – –
Extensor digiti quinti proprius	– – –	x x x	– – –	– – –
Extensor carpi ulnaris	– – –	x x x	– – –	– – –
Supinator	x x x	– – –	– – –	– – –
Abductor pollicis longus	– – –	– – –	x x x	– – –
Extensor pollicis brevis	– – –	– – –	x x x	– – –
Extensor pollicis longus	– – –	– – –	x x x	– – –
Extensor indicis proprius	– – –	– – –	x x x	– – –
Median nerve—C6–7–8, T1				
Flexor digitorum profundus–1 and 2	x x x	– – –	– – –	x x x
Pronator teres	– – –	– – –	– – –	x x x
Palmaris longus	x x x	– – –	– – –	x x x
Flexor carpi radialis	x x x	– – –	– – –	– – –
Flexor digitorum superficialis	x x x	– – –	– – –	x x x
Flexor pollicis longus	x x x	– – –	– – –	– – –
Pronator quadratus	– – –	x x x	– – –	– – –
Abductor pollicis brevis	– – –	x x x	– – –	– – –
Opponens pollicis	– – –	– – –	– – –	x x x
Flexor pollicis brevis	x x x	– – –	– – –	– – –
Lumbricales 1 and 2	Mass closing of hand	Mass opening of hand	Mass opening of hand	Mass closing of hand
Ulnar nerve—C8, T1				
Flexor carpi ulnaris	– – –	– – –	– – –	x x x
Flexor digitorum profundus–3 and 4	x x x	– – –	– – –	x x x
Flexor pollicis brevis	x x x	– – –	– – –	x x x
Palmaris brevis	– – –	– – –	– – –	x x x
Abductor digiti quinti	– – –	x x x	– – –	– – –
Opponens digiti quinti	x x x	– – –	– – –	– – –
Flexor digiti quinti	x x x	– – –	– – –	– – –
Dorsal interossei	– – –	x x x	x x x	– – –
Palmar interossei	x x x	– – –	– – –	x x x
Adductores pollicis	x x x			
Lumbricales 3 and 4	Mass closing of hand	Mass opening of hand	Mass opening of hand	Mass closing of hand

Note: Involvement of triceps is indication for performing patterns with elbow straight when distal components are emphasized.

Note: Distal components are more easily controlled if the elbow remains straight although two-joint action of muscles may be considered. Maximal re-education requires all combinations of elbow joint motion.

Optimal patterns according to peripheral innervation (Tables 14-15)

Table 15. Optimal Patterns for Lower Extremity Muscles According to Peripheral Innervation

Nerve	Flexion–adduction–external rotation	Extension–abduction–internal rotation	Flexion–abduction–internal rotation	Extension–adduction–external rotation
L1–2–3				
Psoas minor	x x x	– – –	– – –	– – –
Psoas major	x x x	– – –	– – –	– – –
Femoral, L2–3–4				
Iliacus	x x x	– – –	– – –	– – –
Pectineus	x x x	– – –	– – –	– – –
Sartorius	x x x Knee flexion	– – –	– – –	– – –
Rectus femoris	x x x Knee extension	– – –	x x x Knee extension	– – –
Vastus medialis	x x x Knee extension	– – –	– – –	x x x Knee extension
Vastus lateralis	– – –	x x x Knee extension	x x x Knee extension	– – –
Vastus intermedius	– – –	x x x Knee extension	x x x Knee extension	– – –
Articularis genu	x x x Knee extension	x x x Knee extension	x x x Knee extension	x x x Knee extension
Obturator nerve, L3–4				
Obturator externus	x x x	– – –	– – –	– – –
Adductor magnus	– – –	– – –	– – –	x x x
Adductor longus	x x x	– – –	– – –	– – –
Adductor brevis	x x x	– – –	– – –	– – –
Gracilis	x x x Knee flexion	– – –	– – –	– – –
Superior gluteal L4–5, S1				
Gluteus medius	– – –	x x x	– – –	– – –
Gluteus minimus	– – –	x x x	– – –	– – –
Tensor fasciae latae	– – –	– – –	x x x	– – –
L5, S 1				
Quadratus femoris	– – –	– – –	– – –	x x x
Gemellus inferior	– – –	– – –	– – –	x x x
S1,2,3				
Obturator internus	– – –	– – –	– – –	x x x
Gemellus superior	– – –	– – –	– – –	x x x
S1,2				
Piriformis	– – –	– – –	– – –	x x x
Inferior gluteal L5, S1–2				
Gluteus maximus	– – –	– – –	– – –	x x x
Sciatic nerve, L4,5; S1,2,3				
Semitendinosus	x x x Knee flexion	– – –	– – –	x x x Knee flexion
Semimembranosus	x x x Knee flexion	– – –	– – –	x x x Knee flexion
Biceps femoris— long & short heads	– – –	x x x Knee flexion	x x x Knee flexion	– – –
Common peroneal nerve L4–5; S1–2				
Tibialis anterior	x x x	– – –	– – –	– – –
Extensor digitorum longus	x x x	– – –	x x x	– – –
Extensor hallucis longus	x x x	– – –	x x x	– – –
Peroneus longus	– – –	x x x	– – –	– – –
Peroneus brevis	– – –	– – –	x x x	– – –
Extensor digitorum brevis	x x x	– – –	x x x	– – –
Peroneus tertius	– – –	– – –	x x x	– – –
Tibial nerve, L4–5, S1–2				
Gastrocnemius	– – –	x x x Knee flexion	– – –	x x x Knee flexion
Popliteus	– – –	x x x Knee flexion	x x x Knee flexion	– – –
Plantaris	– – –	– – –	– – –	x x x Knee flexion
Soleus	– – –	x x x	– – –	x x x
Tibialis posterior	– – –	– – –	– – –	x x x
Flexor digitorum longus	– – –	x x x	– – –	x x x
Flexor hallucis longus	– – –	x x x	– – –	x x x

Reference tables

Table 15. Optimal Patterns for Lower Extremity Muscles According to Peripheral Innervation (Continued)

Nerve	Flexion–adduction–external rotation	Extension–abduction–internal rotation	Flexion–abduction–internal rotation	Extension–adduction–external rotation
Medial plantar nerve L5–S1				
Flexor digitorum brevis	– – –	x x x	– – –	x x x
Abductor hallucis	x x x	– – –	– – –	– – –
1st Lumbrical	x x x	x x x	x x x	x x x
Flexor hallucis brevis	– – –	x x x	– – –	x x x
Lateral plantar nerve L5–S1–2				
Quadratus plantae	– – –	x x x	– – –	x x x
Abductor digiti quinti	– – –	– – –	x x x	– – –
Flexor digiti quinti brevis	– – –	x x x	– – –	– – –
Opponens digiti quinti	– – –	x x x	– – –	– – –
Adductors hallucis	– – –	x x x	– – –	– – –
Plantar interossei	– – –	x x x	– – –	x x x
Dorsal interossei	x x x	– – –	x x x	– – –
3 Lateral lumbricales	x x x	x x x	x x x	x x x

Note: Distal parts may be more easily controlled if the knee remains straight. Unless flexion or extension of the knee are specified, any combination of knee motion may be used.

Note: Gray, Henry. "Anatomy of the Human Body," 23rd ed. Lea & Febiger, Philadelphia, 1936 was reference used in construction of Peripheral Innervation Tables.

Index

Movement (*continued*)
    development of direction, 112
    dexterity and speed of, 11
    direction of eye, 187
    factors which limit, 105
    initiation of
        in ranges of total patterns, 119
        technique for, 95
    mouth, related patterns, 187
    rate of, repetition and, 116
    reversal, in total pattern, 112
    speed of, maximal resistance and, 87
    tongue, 187
    total patterns of
        combining components, 112
        directions used, 122
        and posture, goals in performance, 122
Muscle(s)
    co-contraction of, 97, 162
    components, major
        lengthened range, 12
        optimal contraction of, 12, 13
        overlapping of, 11
        topographical relationships, 11
        versatility of, 11
    contraction
        transitions between types of, 118
        types of, 14
    perineal, stimulation of, 188
    synergy of, 13
    topographical relationships in reinforcing patterns, 91

Neuromuscular mechanism
    deficient, 4
        use of developmental activities, 90
    normal, 4
Nomina Anatomica, 5
Normal subjects
    fatigue in, 92
    "groove" of patterns, 12
    mass movement in, 9
    normal timing, 13, 88
    performance of, defeating, 193
    pressure as facilitating mechanism, 84
    reinforcement in, 89
    rhythmic stabilization in, 97
    variations among, 193

Overflow. *see* Irradiation

Pain
    avoidance of, 84
        by rhythmic stabilization, 97
        in use of stretch, 86
    cold for relief of, 105
    muscle spasm and use of hold-relax, 99
    presence of
        and manual contacts, 84
        and tone of voice, 85
Parkinsonism, 95
Passive motion, in rhythmic initiation, 95
Passive stretching, substitutes for, 98

Pattern(s)
    agonistic, defined, 13
    antagonistic
        as range-limiting factor, 13
        defined, 13
        "groove" of, 12
        shortening and lengthening reactions, 13
    combining, used in wheeling chair, 173
    component, forms of combinations, 112
    muscle components, topographical relationships, 11
    of facilitation
        combining, technique of, 89
        diagonal characteristic, 12
        extremities
            consistency of motion components, 10
            naming of, 10
        positioning for performance, 14
        positioning of part, 12
        range of initiation, 12
        range, parts of, defined, 12
        spiral characteristic, 12
        thrusting variations, 54
Pavlov, I., 3
Pivots of emphasis, shifting manual contacts, 84
Positioning
    for performance, 14
    of therapist, 84
    for use of stretch, 85
Posture, erect, diagonals in total pattern, 162
Practice, automatic, 111
Pressure
    applied during respiration, 185
    for facilitation, 84
    see also Manual contacts
Principles of method, 4
Prone locomotion, thrusting patterns and, 54

Range of motion, evaluation of
    active, 195
    passive, 194
Range of patterns, defined, 12
Range-limiting factors, 13
Recapitulation, of developmental activities, in stress situations, 115
Receptors, joint, techniques directed toward, 86
Reciprocal innervation, in facilitation techniques, 98
Recuperation, change of activity, conversation, 85
Reflexes
    asymmetric tonic neck, 111
    considered in positioning, 14
    gag reflex, 188
    hyperactive, correction of imbalances, 95
    Moro, as total response, 111
    postural
        employed, 3
        stimulation of, 86
    postural and righting, in standing balance, 162
    reinforcement by, 89

Reflexes (*continued*)
    stretch, use of, 93
    tonic neck, and hand-eye coordination, 113
Reinforcement
    automaticity of, in stress situations, 89
    combining patterns, other techniques used, 91
    by reflexes, 89
    see also Irradiation
Relaxation
    voluntary, in rhythmic initiation, 95
    see also Inhibition
Relaxation techniques
    contract-relax, 98
    hold-relax, 99
    slow reversal-hold-relax, 99
Repeated contractions, repetitions controlled in retrograde, 119
Repetition
    and endurance, 87, 93
    of motion, 87
    see also Practice
Resistance
    grading of, 87
    maximal, defined, 87
Response(s)
    developing, auditory, 113
    facilitated, evaluation of, 195
    primitive, in stress situations, 115
    righting, 112
    selectivity of, 116
    to stress, 115
    total, of fetus, 111
    withdrawal, from painful pressure, 84
Reversal of antagonists
    techniques
        rhythmic stabilization, 97
        slow reversal, slow reversal-hold, 96
Rhythmic stabilization technique, 97
Rood, M. S., 105

Safety, of mat activities, 119
Sensory cues
    pressure, 84
    tapping, 84
    tone of voice, 85
    vision, 14
        and direction, 85
        and reinforcement, 89
Sequence, developmental, recapitulation of, 115
Sherrington, C., 3, 4, 9, 95, 98, 118
Shifting of emphasis
    recuperative motion, 92
Skin contact, and communication, 84
Slow reversal-hold-relax technique, 99
Slow reversal-hold technique, 96
Slow reversal technique, 96
Soft palate, stimulation of, 188
Stimulation
    electrical, as adjunct to techniques, 106
    by reversal of antagonists, 96
    sensory in fetus, 111
    soft palate, 188
    see also Facilitation